BEEKEEPER'S LAB

BEEKEEPER'S LAB

52 FAMILY-FRIENDLY ACTIVITIES AND EXPERIMENTS
EXPLORING THE LIFE OF THE HIVE

KIM LEHMAN

QUARRY

Brimming with creative inspiration, how-to projects, and useful information to enrich your everyday life, Quarto Knows is a favorite destination for those pursuing their interests and passions. Visit our site and dig deeper with our books into your area of interest: Quarto Creates, Quarto Cooks, Quarto Homes, Quarto Lives, Quarto Drives, Quarto Explores, Quarto Gifts, or Quarto Kids.

First Published in 2017 by Quarry Books, an imprint of The Quarto Group, 100 Cummings Center, Suite 265-D, Beverly, MA 01915, USA.
T (978) 282-9590 F (978) 283-2742
QuartoKnows.com

Quarry Books titles are also available at discount for retail, wholesale, promotional, and bulk purchase. For details, contact the Special Sales Manager by email at specialsales@quarto.com or by mail at The Quarto Group, Attn: Special Sales Manager, 100 Cummings Center, Suite 265-D, Beverly, MA 01915, USA.

ISBN: 978-1-63159-268-3

Digital edition published in 2017

Library of Congress Cataloging-in-Publication Data
Names: Lehman, Kim (Beekeeper)
Title: Beekeeper's lab : 52 family-friendly activities and experiments
   exploring the life of the hive / Kim Lehman.
Other titles: Bee keeper's lab
Description: Beverly, Massachusetts : Quarry Books, an imprint of Quarto
   Publishing Group USA, Inc., [2017]
Identifiers: LCCN 2016043488 | ISBN 9781631592683 (flexi-bind)
Subjects: LCSH: Bee culture. | Bee products. | Honey. | Beeswax.
Classification: LCC SF523 .L44 2017 | DDC 638/.1--dc23
LC record available at https://lccn.loc.gov/2016043488

Design and Page Layout: *tabula rasa* graphic design

See page 140 for photo credits

IN MEMORY OF MY MENTOR AND DEAR FRIEND, MACK RAY.

CONTENTS

INTRODUCTION

"There are certain pursuits which, if not wholly poetic and true, do at least suggest a nobler and finer relation to nature than we know. The keeping of bees, for instance."

—Henry David Thoreau

I was a free-range child. Each summer my brothers, cousins, and I spent the days gallivanting around our little valley of Soap Hollow in western Pennsylvania with flower chains around our necks, mouths stained with blackberries, and bare feet in constant contact with the earth.

The delight and wonder of those days has inspired a life filled with gardening and herbs. With my deep connection to nature, it was only a matter of time before honeybees entered my world. One captured swarm was all it took to get me hooked on honeybees.

The connection between honeybees and humans goes back thousands of years. We have had generations to learn about these tiny creatures yet, even with modern technology and research capabilities, many aspects of a bee's life remain a mystery. It is strangely comforting to know that bees still have secrets. Although we don't fully understand all the workings of the hive, one thing is certain: Human beings are inextricably linked to bees.

These tiny creatures help make a healthier, sweeter, and more beautiful world. Some of the most nutritious foods in our diet depend on insect pollination. Honey provides delicious treats and natural home remedies. Pollen and propolis can be made into tinctures, extracts, and healing salves. Beeswax, elegant and useful, is made into art, candles, and lotions.

Let's celebrate our connection to the fascinating and humble honeybee through simple experiments, beeswax creations, art, activities, and outdoor adventures. Whether you are a beekeeper or honeybee advocate, I hope the labs in this book will bring you and the children in your life a greater connection to honeybees, beekeeping, and the many products created in partnership with bees and plants that provide us with joy, beauty, and health.

Explore. Make. Create. Protect.

OVERVIEW

Honeybees have the starring role in the varied projects featured throughout this book. Whether finding the queen in a hive, making herb-infused honey, creating a pollination journal, or helping bees by planting a bee-friendly garden, the honeybee is ever present.

The labs in this book contain a list of materials, step-by-step processes, ways to extend the activities, suggestions for young children, and facts about honeybees. Additionally, some labs include background information and content to increase knowledge and provide a better understanding of the activity.

Each lab is a stand-alone project, but in many cases labs can complement and enhance each other. For instance, beeswax art blocks (Lab 44) can be used with encaustic cards (Lab 45) and comb rubbings (Lab 43).

Some of the supplies needed for the labs are the products of the hive: honey, pollen, propolis, and beeswax. If you are a beekeeper, you can collect, gather, scrape, and extract these materials from your hive, taking great care to use what you take. If you are not a beekeeper, these products can be purchased from a local beekeeper or a health food store or ordered online.

May the world of bees and beekeeping continue to open many doors to explorations, curiosity, and continued learning for you and your family.

Let's celebrate honeybees!

POLLEN COLLECTION

Produced in the anther of a flower, pollen grains carry the male reproductive cells of the plant. Pollen provides developing bee larvae with a vital food source that is rich in protein and other essential nutrients.

• A pollen trap placed at the hive entrance makes it possible to collect pollen from the legs of foraging bees as they enter the hive. When bees squeeze through the wire mesh of the trap, about 50 percent of the pollen is knocked off their legs and falls into a tray or holding compartment.

• To ensure the bees are delivering enough pollen for the brood, inactivate or remove pollen trays for a few days every week while the bees are collecting pollen. Remove pollen trays at all other times.

• Collect pollen from the trays daily, if possible, or at least every few days. Daily collections will be necessary in high-humidity areas or during times of inclement weather because moisture may lead to mold growth. Briefly blow dry air over the pollen with a hair dryer to dry slightly.

• Pollen is perishable, so care in storing is vital. Avoid exposing pollen to high heat and direct sunlight. Store pollen in the refrigerator or freezer.

• Use collected pollen to feed back to the bees.

PROPOLIS COLLECTION

Plants produce propolis, a sticky resin, to protect buds and wounds from bacteria and fungus. Bees gather propolis from plants and trees such as alders, birch, conifers, poplar, and willows and then use it to seal their hive, embalm intruders, and coat the brood cells.

• A special propolis trap may be purchased and used in place of the hive's inner cover.

Propolis

• After removing the propolis trap, place it in a plastic bag and freeze. Knock the propolis off the trap by using sudden force or twisting or scraping it with a hive tool.

• Propolis can be scraped and collected from frames, lids, and hive boxes.

• If using hive scrapings, freeze immediately in plastic bags to make cleaning debris easier.

• Spread the scrapings on a cookie sheet and manually pick out wood chips. Another cleaning option is to place the propolis in a basin of water, agitate the basin, and the wood pieces will float to the top.

• Propolis is sticky, so use only utensils that are expendable.

BEESWAX TIPS

Beeswax is produced from four pairs of glands on the underside of a worker bee's abdomen. The wax is used for building comb in the hive.

• Beeswax has a melting point of around 145°F (62.8°C).

• Make cleanup easier by protecting your work surface with newspaper. A layer of cardboard on the floor is also recommended.

• A simple double boiler can be made with a can or small pot in a larger pot of hot water to safely melt wax.

• For a final filter, pour melted wax through a paper towel or sweatshirt material.

• Prolonged exposure to temperatures of 180°F (82°C) and higher will darken the wax.

BEESWAX SAFETY

• Never leave hot wax unattended.

• Beeswax is highly flammable. Keep a fire extinguisher within reach.

• Always use an electric heating element. No open flames.

• Unplug equipment when not in use.

Beeswax

UNIT 1

BEEKEEPING

HONEYBEES TEACH US THAT PATIENCE IS A PRIORITY.

When you open a hive, honeybees quickly inform you that it's time to slow down. Every movement and every action needs to be calm and deliberate. The outside world disappears. The hectic pace of everyday life has no place in a humming hive. Slow down. Be patient.

Simply sitting quietly by the hive entrance observing the comings and goings of the bees holds an almost meditative quality—the movement, the sounds, just being in the presence of these noble creatures draws me in.

Honeybee stewardship requires much work and responsibility, but rewards are found not only in the honey itself but also in the elation of finding a queen or in the delight of chewing on comb honey you worked with the bees to produce. In this unit, we will explore some of the joys and responsibilities of keeping bees.

Search. Do. Construct. Harvest.

SETTING UP A HIVE

YOU WILL NEED

- 2 deep hive bodies and frames
- 20 sheets of foundation
- 1 or 2 medium or shallow supers and frames
- 10 to 20 sheets of medium or shallow foundation
- inner and outer covers
- bottom board
- queen excluder
- feeder
- veil and bee suit
- gloves
- smoker and fuel
- hive tool
- hive stand
- honeybees

BEE BUZZ

For every 100 beekeepers, about 95 percent are hobbyists, 4 percent are sideliners, and 1 percent are full-time or commercial beekeepers. Generally speaking, hobbyists have a few hives and sideliners seek to make a profit but also have another income source.

There are many reasons people are interested in keeping bees: honey production, garden or orchard pollination, support of the honeybee population, a hobby for retirement, or just the opportunity to be a part of something vital and extremely interesting. This lab offers a brief overview of things to consider when setting up a hive.

DIRECTIONS

1. **Read.** Educate yourself about bees and beekeeping as much as possible.

2. **Join a beekeeping club.** The camaraderie of a group of beekeepers offers support, information, and knowledge when you need it the most.

3. **Hive equipment.** The Langstroth hive is the standard equipment used by most beekeepers. Each box or super is filled with movable frames and foundation to help manipulate the hive. Have all your equipment prepared well in advance of the estimated arrival of your packaged bees or nucs. (Figs. 1 and 2)

4. **Safety gear.** You will need a full suit or a jacket and veil along with leather, canvas, or latex gloves.

5. **Tools.** Always use a smoker when opening a hive. You will also need fuel for the smoker (Lab 2), some way to feed the bees (Lab 4), and a hive tool to separate hive equipment stuck together with propolis. (Fig. 3)

6. **Apiary site.** Help guide the flight paths of bees, especially in urban areas, up and away from human activity. A fence, tall plants, or the side of a building can help accomplish this.

7. **Hive orientation.** Two hives on one stand with 8' (244 cm) of separation between the pairs of hives works well. Avoid placing hives in rows of six or more. These long rows can cause bee drifting, the spread of diseases, and the robbing of weaker colonies. A better setup is a U shape with the entrances facing toward the middle. Having hive entrances face in different directions can cut down on drifting as well.

8. **Hive stand.** A sturdy hive stand is essential. Repurposed campaign signs made of corrugated plastic placed in front of the hive, though not pretty, provide a great weed barrier and prevent drifting by offering bees a visual cue to their home. (Fig. 4)

9. **Bees.** Purchase packaged bees, nucs (three to five frame colonies), or an established hive from a local beekeeper or bee producer. Place an order for bees in December or January to ensure availability and timely delivery. Of course, catching wild swarms is another option for acquiring bees.

PARTS OF A BEEHIVE

Outer cover: Every house needs a roof. This cover protects the hive from the elements.

Inner cover: This cover offers some summer heat protection, helps with airflow, and provides a platform to feed bees in the hive.

Honey supers: These medium and shallow boxes are where the worker bees make and store surplus honey. As the bees fill the frames in each box with honey, another super can be added to the hive.

Queen excluder: This metal or plastic rack is placed between the brood chamber and the honey supers. Its main purpose is to restrict the queen from moving into the honey supers to lay eggs. Not all beekeepers use a queen excluder.

Hive body or brood chamber: You will need two hive boxes or three medium-size supers full of frames and foundation to provide the space needed for comb building, egg laying, brood development, and pollen stores. Some supers hold eight frames; others hold nine or ten frames.

Frames and foundation: Frames come in many sizes and styles. Each frame holds wax or plastic foundation embossed with the shape of worker cells. The foundation helps with hive manipulation and honey extraction. (See fig. 2 on page 19.)

Bottom board: This is the floor of the hive. It can be made of solid wood or screened mesh.

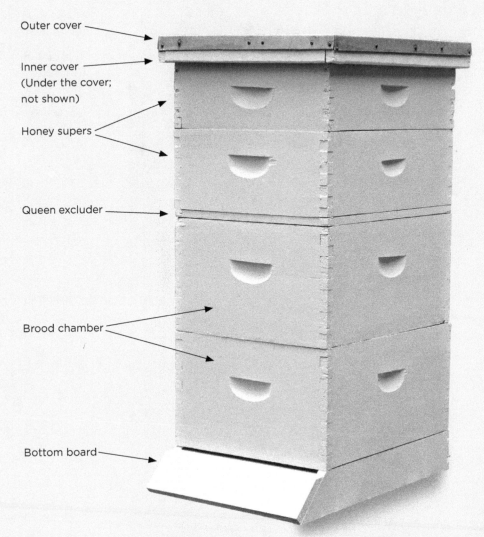

Outer cover

Inner cover
(Under the cover; not shown)

Honey supers

Queen excluder

Brood chamber

Bottom board

Fig. 1: *Parts of a beehive.*

APIARY SITE AND HIVE PLACEMENT

Here is a quick checklist of things to consider when choosing a hive location:

- If neighbors are close by, consider using fences or plants to help direct the flight paths of the bees and obscure the hive from view.

- Make sure there is plenty of room to move completely around the hive.

- The apiary needs to be in a fairly convenient location to ensure frequent visits.

- Vehicle accessibility eases the need to carry equipment long distances.

- Water drainage away from the hive will help avoid soggy areas during heavy rains.

- A nearby water source, either natural or provided, offers efficient water collection for the bees. It also keeps neighbors happy by reducing bee visits to pools and pet watering bowls. (Fig. 5)

- Nectar- and pollen-producing plants are essential for healthy, happy bees.

 TAKE IT FURTHER

- Once you have at least two years of beekeeping under your belt, offer to mentor a new beekeeper. You don't have to feel like an expert to help someone who knows less than you!

- Try your hand at beekeeping using a top bar hive. The equipment required is inexpensive, there is no need for equipment storage space, lifting is minimal, and bees construct their own comb without the use of manmade foundation.

- Learn more about what beekeepers refer to as "bee space."

Fig. 2: *Each super holds eight to ten frames with plastic or wax foundation.*

Fig. 4: *A simple hive stand can be made using cinder blocks and thick boards. One way to decrease weed growth at the hive entrance is by mulching with wood chips or rocks.*

Fig. 3: *Two important tools every beekeeper needs: a hive tool and a smoker.*

Fig. 5: *Prevent bees from being a nuisance to neighbors by providing a steady water source. On average, in the summer, a hive of bees uses a quart (1 L) or more of water daily.*

SMOKER BUNDLES

YOU WILL NEED

- scissors
- twine
- pine needles

BEE BUZZ

Smoke masks the alarm pheromones bees release when they believe there is a threat to the hive. Also, in nature, smoke signifies a wildfire, warning the bees they may have to leave their home. Preparing for the move, they gorge themselves with honey.

Knowing how to light a smoker and keep it lit is an essential skill for successful beekeeping. These bundles of fuel are easy to make, compact, and efficient.

DIRECTIONS

1. Cut a piece of twine about the length of your arm.

2. Gather a handful of pine needles. Use both hands to fold long pine needles once or twice to form a bundle. (Fig. 1)

3. Grasp the pine needle bundle with one hand. Using the other hand, slip the end of the twine under the thumb of the hand holding the bundle. (Fig. 2)

4. Bind the pine needles together by continuing to wrap the twine down the length of the bundle. (Fig. 3)

5. Tie the ends of the twine together to secure.

FUN FOR KIDS

Younger children can contribute to the process by gathering fuel such as pine needles, leaves, and bark.

TAKE IT FURTHER

- Why not have some friendly smoker-lighting competitions? Who can keep the smoker going the longest? Each person has five minutes to light the smoker. Once the smoker is lit, two puffs are allowed on the bellows every five minutes. Know that it is possible to keep the smoker burning for a very long time. You may want to plan a picnic as part of this activity.

- Complement these pine needle bundles by making fire starters. Dip pinecones in 100 percent beeswax or make fire starters from newspaper (Lab 23).

Fig. 1: *Form a bundle of pine needles using both hands.*

Fig. 2: *Slip one end of the twine under the thumb holding the pine needle bundle.*

Fig. 3: *Wrap the twine tightly around the pine needle bundle.*

USING SMOKER BUNDLES

1. Begin with an empty smoker.

2. Light a piece of crumpled newspaper. Use your hive tool to push the burning paper down into the smoker.

3. While the flames are visible, place the pine needle bundle in the smoker.

4. Puff the bellows a few times to keep air moving through the smoker.

5. One bundle will last about an hour. You can always add another bundle to keep the smoker going.

6. Maintain the smoldering fire by intermittently puffing the bellows.

7. When finished, plug the nozzle to put out the fire or lay the smoker on its side. Do not dump the smoker contents in a dry area. Make sure the fire in the smoker is completely out before storing.

SMOKER FUEL

Many smoker materials are widely available. Make sure the fuel is untreated and free of chemicals, paint, and plastics.

- pine needles
- pinecones
- dry grass
- dry leaves
- bark
- punky wood
- wood shavings
- animal bedding
- burlap, untreated
- twine, untreated

WAXING FOUNDATION

YOU WILL NEED

- beeswax
- slow cooker or double boiler
- foam brush
- plastic foundation in a frame

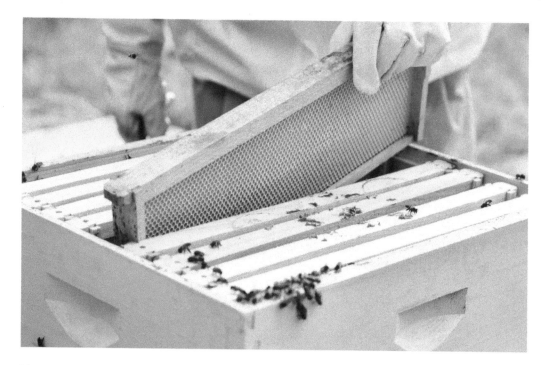

BEE BUZZ

When honeybees build their own comb naturally without foundation, comb thickness and the orientation of the cells can vary.

Plastic foundation has a number of advantages over other types of foundation. It is sturdy, comparatively inexpensive, and fast to install. The foundation can be scraped, power washed, rewaxed, and reused. Bees prefer plastic foundation that is coated with a thin layer of beeswax. The goal is to apply wax to the tops of the formed cell shapes in the foundation. The wax needs to be hot enough to avoid clumping on the brush but not so hot that it fills the base of the cells. A little experimenting will do the trick. It helps to keep the brush warm.

DIRECTIONS

1. Melt the beeswax in a slow cooker or double boiler. No need to use your best wax cappings for this process.

2. Dip the foam brush into the melted wax. (Fig. 1)

3. Brush lightly back and forth over the plastic foundation. (Fig. 2)

4. Repeat on the other side.

Fig. 1: *Cover the tip of the foam brush with wax.*

Fig. 2: *Apply the melted wax to the plastic foundation using back-and-forth brushstrokes.*

FUN FOR KIDS

- Instead of wax, brush tempera paint onto the plastic foundation. Press paper onto the paint to make prints.

- Improve hand-eye coordination by using an eyedropper to fill the indentations on plastic foundation with water to represent a bee proboscis delivering nectar to the hive.

 # TAKE IT FURTHER

Experiment with different types of foundation and frames in your hive. What do your bees prefer? Are the bees consistent with the comb building on each type of foundation? Is there really a difference between waxed and unwaxed plastic foundation? Try placing one frame without foundation in your beehive to observe and compare the natural comb-building tendencies of honeybees.

MAKE A TOP FEEDER

YOU WILL NEED

- handsaw
- 1" × 2" (2.5 × 5 cm) board, 2' (60 cm) long
- marker
- jar with lid
- drill with a $\frac{3}{8}$" (1 cm) bit
- jigsaw
- 8 (1½"; or 3.8 cm) screws or nails
- hammer
- 6" × 7¼" (15 × 18.5 cm) plywood, at least ½" (1.3 cm) thick
- sugar syrup

BEE BUZZ

Depending on the geographic location, a hive will need between 40 and 100 pounds (18.1 and 45.4 kg) of stored honey to survive the winter.

Sometimes bees need supplemental feeding, especially in the spring to promote comb building for new hives or in the fall to build up winter stores. This is just one of the many ways to feed bees.

DIRECTIONS

1. With a handsaw, cut the 1" × 2" (2.5 × 5 cm) board into four 6" (15 cm) sections.

2. Trace around a jar lid with a marker to mark a cutting line in the center of the plywood. (Fig. 1)

3. Drill a hole on the line big enough to insert a jigsaw blade. Cut along the drawn line using the jigsaw. (Fig. 2)

4. Screw or nail one of the 6" (15 cm) pieces of board along the 6" (15 cm) side of the plywood. Support the plywood with another piece of board. (Fig. 3)

5. Screw or nail the second 6" (15 cm) board along the other 6" (15 cm) side of the plywood.

6. Attach one 6" (15 cm) piece of board along one 7¼" (18.5 cm) side. Attach the last 6" (15 cm) board to complete the perimeter. (Fig. 4)

7. Using a hammer and nail, punch a few small holes in the lid of the jar for the bees to drink from.

8. Place the feeder and a jar of sugar syrup on top of the inner cover in a hive. Add an empty super and hive top.

Note: Sugar syrup for the spring is a 1:1 ratio of sugar to water. Autumn feed is generally thicker, with a 2:1 ratio of sugar to water.

TAKE IT FURTHER

- Keep records of the feeding dates and the amount of sugar syrup consumed by each hive.

- Expand this feeder to include holes for two jars.

- Explore other ways of feeding bees, such as Boardman feeders, division board feeders, dry sugar, fondant, pails, or plastic bags.

Fig. 1: *Mark a cutting line on a small piece of plywood by tracing around the jar lid.*

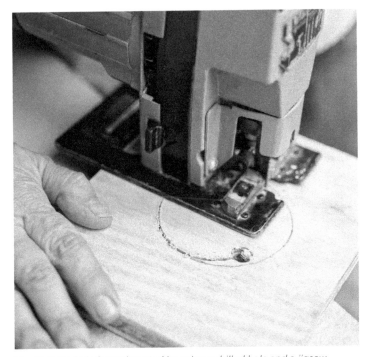

Fig. 2: *Cut a hole for an inverted jar using a drilled hole and a jigsaw.*

Fig. 3: *Attach one side of the feeder box to the plywood using a drill.*

Fig. 4: *Continue screwing the 6" (15 cm) boards around the plywood perimeter.*

FUN FOR KIDS

Make a feeding station outside by pouring sugar syrup onto a plate. Make sure there are places for the bee to stand and not drown, such as on a flat rim, rocks, or a paper towel. Use a magnifying glass to take a closer look at how a bee uses its proboscis to collect food.

DRONE PETTING

YOU WILL NEED

- a hive of bees
- protective beekeeping gear (suit, veil, gloves)
- smoker and fuel
- matches
- hive tool
- honey in a squeeze bottle
- magnifying glass (optional)

Increase you and your family's comfort level around bees by getting up close and personal with drones. This activity also helps when learning to recognize the difference between the queen and drone bees.

DIRECTIONS

1. Open an active hive of bees in the spring or summer by donning protective gear and puffing a little smoke at the hive entrance.

2. With a hive tool, pull out frames one by one to look for drones. They will be bigger than the worker bees.

3. When spotted, carefully pick up the drone by the wings with your thumb and pointer finger or lightly grasp around the thorax. (Fig. 1)

4. Depending on your comfort level and bee activity, move to a bee-free zone or take off your gloves near the hive. Squeeze a little drop of honey onto a bare hand so the drone will stick around. No pun intended. (Fig. 2)

5. Examine the drone using a handheld magnifier.

6. When done, return the drone to the hive.

BEE BUZZ

The number of drones in a hive varies from zero in the winter to thousands in the spring and summer. Drones are usually about 15 percent of the total bee population in a hive.

FUN FOR KIDS

- What are the differences and similarities between a drone and a worker bee?

- Find both drone and worker brood in the hive.

Fig. 1: *Carefully pick up drones with ungloved hands.*

Fig. 2: *Stick the drone to a drop of honey on your hand for closer examination.*

ABOUT DRONE BEES

• A drone is a male honeybee.

• He does not have a stinger.

• A drone's abdomen is more rounded than the queen's.

• The eyes are large and can be seen on the top of the head.

• These bees are hatched from unfertilized eggs laid by the queen in drone cells.

• Their preferred flying time is in the midafternoon, unless they are leaving with a swarm in the morning.

TAKE IT FURTHER

• Did you know that drones can juggle? Pick up a drone bee by the wings. Hold the bee upside down with his legs in the air. Place a small rounded piece of cork on the legs and watch the bee juggle for a few seconds. Thank the drone for sharing its hidden talents! Return the drone to the hive by placing it on the landing board at the hive entrance.

• Drones can make wonderful bee ambassadors at public events. Collect about fifty drones on the day of the event. For easy retrieval, carry the drones in a butterfly house along with a small sponge of sugar water. Put a small drop of liquid chocolate on your finger; it is not as drippy or sticky as honey. Place a drone on top of the chocolate. Ask folks if they would like to pet your bee. Now that you have their attention, take the opportunity to answer questions and share important information about honeybees.

FIND THE QUEEN

YOU WILL NEED

- a hive of bees
- beekeeping protective gear (suit, veil, gloves)
- smoker and fuel
- matches
- hive tool

No matter how long you have been keeping bees, excitement mounts when the queen bee is spotted using this procedure.

DIRECTIONS

1. Open an active hive of bees in the spring or summer by donning protective gear and puffing a little smoke under the hive cover. You want the queen to move down into the hive.

2. With a hive tool, remove any honey supers.

3. If you have two deep brood boxes, separate the two by placing the top one on an upside-down hive cover on the ground. Begin the queen search in the lower brood box.

4. If possible, stand on the side of the hive with your back to the sun so the frames are in the light. Begin by carefully pulling out and examining the frame closest to you. Scan one side of the frame and then slowly turn it over to examine the other side. A good way to do this is by moving the frame to a vertical position. Slowly spin the frame around and then move the frame back to a horizontal position for examination. Lean this frame outside the hive to create space in the super.

5. Remove the next frame. Slide into the open space before lifting straight up. Examine both sides starting with the surface closest to you. Place it back in the hive, leaving a gap between the unexamined frames. (Fig. 1)

6. Quickly continue this process until you find the queen or have scanned all the frames in the brood nest. (Fig. 2)

7. If the queen is not found, scan the frames in the second brood super.

8. While examining each frame, look for the presence of eggs, larvae, and capped brood. A good solid pattern of capped brood is a sign of a healthy laying queen.

9. If you did not see her, try again another time.

BEE BUZZ

The primary role of the queen bee in the hive is to lay the eggs. On a good day in the summer, she can lay up to 1,500 eggs. She also produces a pheromone that affects the behavior of the bees in the hive.

Fig. 1: *Slide frames over before lifting out the next frame to avoid damaging the queen.*

Fig. 2: *Examine each frame by quickly scanning both sides.*

QUEEN-SEARCHING TIPS

- Use as little smoke as possible.

- Always hold the frames above the brood box in case the queen falls.

- It is unlikely the queen will be on the outermost frames.

- Scan the entire frame quickly. Avoid letting your eyes rest on every bee.

- Observe the actions of worker bees. A healthy queen will have attendants around her.

- The drones and the queen will pop out from the crowd. Familiarize yourself with the physical appearance of a drone (Lab 5).

- Once the queen is found, practice developing a sharper eye by turning your head away from the frame for a few seconds and then looking back to find her again.

- Take care not to crush the queen on the sides of the frame when returning it to the hive.

- Return all frames to their original positions in the hive.

TAKE IT FURTHER

Record your observations in a beekeeping journal. Was there a solid or random brood pattern? Did you see any eggs? Describe the larvae and capped brood.

LAB 7

CUT COMB

YOU WILL NEED

- a strong hive of bees
- thin surplus foundation, shallow or medium
- sharp knife or pizza cutter
- shallow or medium honey supers and wedge-type frames
- construction stapler or small brad nailer with ¾" (2 cm) brads
- marker
- cutting board
- plastic clamshell containers or other packaging boxes
- wire cooling rack or screen
- large cookie sheet
- spatula

BEE BUZZ

A deep frame filled with capped honey will hold 5 to 6 pounds (2.3 to 2.7 kg) of honey. A full shallow frame holds about 3 pounds (1.4 kg).

There is an unexplainable joy in seeing honey in its most natural, unprocessed form directly from the beehive. Begin small by producing one or two frames of cut comb honey for yourself and your family. This process uses foundationless frames for honey production.

DIRECTIONS

1. To gather cut comb from a hive, cut a piece of thin surplus foundation into 1" (2.5 cm) strips using a knife or pizza cutter.

2. Lay the strip along the top ledge of the wedge frame. Staple or nail the wooden wedge over the foundation strip to secure. (Fig. 1)

3. Mark the top of the frame that will be used for cut comb.

4. Insert the cut comb frame in the center of a honey super between two frames of drawn-out comb, if possible.

5. Once the bees have filled the frame with comb and capped honey, remove it from the hive to cut and package the comb honey. Fill the empty space in the hive with a prepared frame.

6. Lay the frame on a cutting board. Use a knife to cut around the perimeter of the frame. (Fig. 2)

7. Cut the comb into the desired shape and size to fit into the plastic packaging containers. (Fig. 3)

8. With a spatula, place the dripping comb on a wire rack atop of a large cookie sheet and let it drain for at least 15 minutes. (Fig. 4)

9. Gently pick up each section with the spatula and place it in a plastic box.

10. Put the boxes of cut comb in the freezer for 2 to 3 days to destroy any unwanted "guests" that may be hanging around.

11. Eat by the spoonful or spread directly on biscuits. You can swallow the wax!

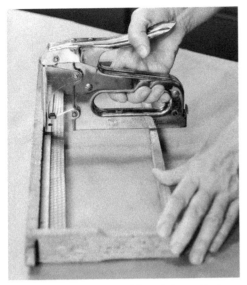

Fig. 1: *Attach a thin strip of foundation to the top of a frame.*

Fig. 2: *Use a sharp knife to cut the comb free from the frame.*

Fig. 3: *Cut the comb into smaller pieces to package.*

Fig. 4: *Allow the honey from the cut edges to drain overnight, if possible.*

TIPS

- Put the frames for cut comb honey on the hive as soon as the spring nectar flow is under way.

- As soon as frames become full, remove them from the hive. Bees will leave little footmarks!

- Use very little smoke when opening the hive to avoid the smell and taste of smoke getting into the honeycomb.

- Dip the knife in hot water and wipe off between cuts.

- Dental floss can be used to cut the comb into squares. Put the floss under the comb, cross the ends on top, and gently pull.

- The cut comb can be stored in the freezer to avoid granulation.

FUN FOR KIDS

Create a multisensory experience by using fingers to break the cappings and scoop out some comb honey. It is finger-licking goodness!

TAKE IT FURTHER

- Expand your cut comb honey production to use as an additional income stream.

- If you have extra pieces of the cut comb, consider making chunk honey. Place a large piece of the comb honey in a wide-mouth glass jar. Fill the jar with light-colored liquid honey.

CRUSH AND STRAIN
HONEY EXTRACTOR

YOU WILL NEED

- 2 food-grade plastic buckets with lids
- permanent marker
- bowl about 1½" (3.8 cm) smaller than the bucket lid
- drill with a ⅜" (1 cm) bit
- jigsaw with a fine-toothed blade at least 17 TPI (teeth per inch [2.5 cm])
- coarse sandpaper
- paint strainer
- capped honey frames
- potato masher (optional)
- plastic wrap

BEE BUZZ

Beekeepers worldwide, along with honeybees, produce about 1.9 million tons (1.7 metric tons) of honey each year.

Make a simple, inexpensive honey extractor out of plastic buckets. For the food-grade buckets, you can buy inexpensive containers used for doughnut filling from a bakery or doughnut shop.

DIRECTIONS

1. Prepare the bottom bucket by cutting a hole in the lid to support the top container. To do this, fasten the lid on the bucket. Mark a cutting line using a permanent marker by placing a bowl on the lid for a template that allows for about 1½" (3.8 cm) around the outside.

2. Drill a hole on the line big enough to insert a jigsaw blade to cut the hole. (Fig. 1)

3. Prepare the top sieve-like bucket. Drill small holes in the bottom of the bucket using a drill with a ⅜" (1 cm), or smaller, bit. The distance between holes should be about two fingers apart. (Fig. 2)

4. Using coarse sandpaper, smooth the edges of the cut lid and drilled holes.

5. Clean the buckets and lids, allowing them to completely dry before extracting your honey.

6. Assemble the extractor. Stretch the paint strainer over the opening of the bottom bucket, pulling the elastic down the outside of the bucket. An alternative strainer can be made with nylon mesh. Place the lid with the cut hole on the lower bucket. Stack the top bucket with the drilled holes on top of the lid. (Fig. 3)

7. Cut (if using top bars) or scrape the capped honey from the frames into the top bucket. Using a potato masher or clean hands, thoroughly crush and squeeze the comb to release the honey. (Fig. 4)

8. Tightly seal the top bucket with the solid lid. Allow the honey to drain and settle for a few days.

9. A thin layer of residue will float to the top of the honey bucket. This can be easily removed by gently pressing a sheet of plastic wrap to the surface of the honey. In one quick motion, swiftly peel the plastic wrap off and out of the bucket into a nearby trash can.

10. The honey is ready to bottle. Save the wax to make candles or cream. Enjoy!

Note: When working with honey, only use plastic, stainless steel, or glass containers.

Fig. 1: *Use a jigsaw to cut the middle out of one plastic lid.*

Fig. 2: *Drill honey drainage holes in the bottom of a plastic bucket.*

Fig. 3: *Assemble the extractor by placing the paint strainer on the bottom bucket, and then the lid with the hole, followed by the bucket with drilled holes.*

Fig. 4: *Crush the honeycomb to free the honey from the sealed wax cells.*

TAKE IT FURTHER

- A gate on the bottom bucket is not absolutely necessary, but it certainly makes bottling the honey easier. You can purchase a bucket with a gate or insert one yourself. Drill a hole about 1½" (3.8 cm) from the bottom of the lower bucket using a hole saw about ⅛" (3 mm) larger than the honey gate. Apply a small amount of lubricant (food-grade grease or olive oil) to the gate's gasket for a tight fit. Insert the gate and tighten with a spud wrench or large pliers.

- Honey varies in color and flavor depending on the nectar sources available to the honeybees. Produce unique artisan honey by extracting several times a year. If you have multiple bee yards, try separating each extraction by location. You can even label your honey by Zip code. After extracting a variety of artisan honey, invite your friends to a honey-tasting party (Lab 14).

- Enter your honey in honey competitions put on by beekeeping organizations. For best results, follow posted rules and regulations exactly.

- Send a sample of your honey to a facility that analyzes the pollen to determine the floral sources in your honey. See the Resources section for more information.

UNIT 2

HONEY

**THE SIMPLE SWEETNESS OF HONEY CONNECTS US
TO THE WORLD OF HONEYBEES AND PLANTS.**

Humankind's long history of desire for honey's sweetness is illustrated in 15,000-year-old cave paintings in Spain depicting the harvest of the prized liquid. The ancient Egyptians revered honey for its medicinal qualities, while Greek philosophers believed honey was an elixir of youth distilled from stars and rainbows. To me, honey is a perfect, generous gift of nature celebrating the special partnership between bees and flowering plants—how sweet it is!

Like many beekeepers, I began keeping bees to harvest honey, a tasty, natural alternative to sugar. I didn't initially realize the incredible amount of labor required by the bees to produce this liquid gold. It is mind-boggling to think that, by some accounts, bees need to travel more than 55,000 miles (88,514 km) visiting nearly two million flowers to produce 1 pound (454 g) of honey!

Surround yourself with sweetness—honey to eat, drink, soothe, and delight. Join with friends in raising a toast in acknowledgment and praise of bees, the industrious workers responsible for every spoonful of honey we enjoy.

Mix. Make. Bake. Taste.

SWEET DOG TREATS

YOU WILL NEED

- ¼ cup (60 ml) vegetable oil
- ¼ cup (65 g) peanut butter
- ½ cup (85 g) honey
- 2 tablespoons (30 ml) water
- 1 egg
- 2½ cups (300 g) whole wheat flour
- ½ cup (40 g) oatmeal
- 1 teaspoon baking powder
- ½ teaspoon ground cinnamon
- baking sheet
- parchment paper
- 2 mixing bowls
- whisk or spoon
- rolling pin
- cookie cutter

Honey is not just for honeybees and humans. Dog friends will thank you for making and sharing these honey and grain tidbits.

DIRECTIONS

1. Preheat your oven to 375°F (190°C or gas mark 5). Line a baking sheet with parchment paper.

2. In a large mixing bowl, combine the oil, peanut butter, honey, water, and egg. Mix well with a whisk or spoon.

3. In a second mixing bowl, combine the flour, oatmeal, baking powder, and cinnamon.

4. Stir the dry ingredients into the oil and peanut butter mixture to make the dough. (Fig. 1)

5. Knead the dough on a lightly floured surface until it comes together into a ball. (Fig. 2)

6. Roll the dough out to about a ¼" (6 mm) thickness. (Fig. 3)

7. Cut out the dough using a cookie cutter dipped in flour. Collect the scraps and lightly knead the dough. Roll out again to cut more shapes. (Fig. 4)

8. Place the cut dough on the prepared baking sheet.

9. Bake for 12 to 14 minutes, until golden and crisp. Cool completely.

10. Store in a sealed container in the refrigerator.

BEE BUZZ

In general, worker bees over 21 days old forage for nectar to make honey.

FUN FOR KIDS

Give the dog owners in your neighborhood these dog treats along with a jar of honey.

Fig. 1: *Combine the flour mixture into the wet ingredients.*

Fig. 2: *Knead the dough to incorporate all the ingredients.*

Fig. 3: *Roll out the dough.*

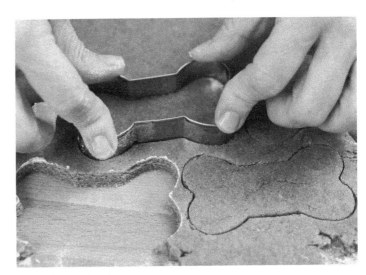
Fig. 4: *Cut out fun shapes.*

 ## TAKE IT FURTHER

Honey for pets is much more than sweet treats. Explore the topical uses of honey for animal wounds, cuts, and burns.

HONEY BUTTER

YOU WILL NEED

- ½ cup (112 g) unsalted butter, at room temperature
- ¼ cup (85 g) honey
- small bowl
- electric mixer
- spoon
- molds
- cutting board
- wax paper

This simple recipe will add an extra special touch to biscuits, muffins, or pancakes.

DIRECTIONS

1. In a small bowl using an electric mixer, whip the butter until smooth. (Fig. 1)

2. Add the honey and mix into the butter. (Fig. 2)

3. Using the back of a spoon, press the butter into molds. (Fig. 3)

4. Freeze or refrigerate until solid.

5. To free from the molds, lay the mold, butter side down, on a cutting board. Slowly run warm water over the mold until the butter is released. (Fig. 4)

6. Place the butter shapes between pieces of wax paper in a sealed container. Keep in the refrigerator or freezer until ready to serve.

VARIATIONS

Using the steps in this lab, try these honey butter variations.

Nutty Butter

- ½ cup (112 g) unsalted butter at room temperature
- ½ cup (130 g) nut butter (almond, peanut, or cashew)
- ¼ cup (85 g) honey
- 2 teaspoons ground cinnamon

Very Berry Butter

- ½ cup (112 g) unsalted butter at room temperature
- 2 tablespoons (40 g) honey
- 2 tablespoons (30 g) berry preserves (blueberry, strawberry, or raspberry)

BEE BUZZ

The European Union (EU), as a region, is the world's largest consumer of honey. The countries within the EU in order of the most honey consumed are Germany, Spain, UK, France, and Italy. In terms of countries, China is the world's largest consumer.

Fig. 1: *Use an electric mixer to whip the butter.*

Fig. 2: *Mix the honey into the butter.*

Fig. 3: *Press the butter into the molds a little at a time to avoid air pockets.*

Fig. 4: *Free the butter from the molds using warm water.*

FUN FOR KIDS

Put the room-temperature honey butter between two pieces of wax paper. Flatten with a rolling pin. Freeze overnight. Cut out shapes using cookie cutters.

 TAKE IT FURTHER

- Package the honey butter in small, wide-mouth jars to give away as gifts.

- Create more honey butter variations using ground cinnamon, ground ginger, vanilla extract, orange zest, or fresh fruit.

LAB 11

HONEY HYDRATOR

YOU WILL NEED

- 3 or 4 lemons
- 4 or 5 limes
- 1 cup (240 ml) coconut water
- 3 tablespoons (60 g) raw, unfiltered honey
- 1 teaspoon bee pollen
- pinch of sea salt
- ½ teaspoon vitamin C powder
- strainer
- canning jar with a tight-fitting lid
- blender (optional)

BEE BUZZ

Honey, a carbohydrate, provides bees with the energy needed to perform their foraging duties in the summer.

⚠ SAFETY NOTE

Be aware that some people may have an allergic reaction to bee pollen.

FUN FOR KIDS

Freeze the honey hydrator to make a cold, healthy treat. After combining the ingredients, pour the liquid into small paper cups. Cover individually with two layers of aluminum foil and place on a cookie sheet. Punch a wooden craft stick through the foil to make a handle. Carefully place on a shelf in the freezer. When solid, peel off the paper cup and thank a bee for this delicious, refreshing treat.

Working in an apiary or a garden during the hot summer months leaves the sweating body depleted of precious electrolytes like sodium and potassium. Replenish your system naturally with honey, an easily digestible energy source, and coconut water, citrus, and salt, which are all rich in electrolytes.

DIRECTIONS

1. Juice the lemons and limes, straining out the seeds. Measure ½ cup (120 ml) of each juice. (Fig. 1)

2. Pour the citrus juices into a canning jar. Add the coconut water, honey, bee pollen, sea salt, and vitamin C powder. As an alternative, substitute beebread (Lab 32) for the honey and eliminate the teaspoon of pollen. (Fig. 2)

3. Seal and shake the jar to incorporate the honey and salt. If using dried pollen, mix in a blender. (Fig. 3)

4. Drink immediately or make larger amounts, refrigerate, and consume within a few days.

Fig. 1: *Juice the lemons and limes.*

Fig. 2: *Add the coconut water to the juice along with the honey, bee pollen, salt, and vitamin C powder.*

 TAKE IT FURTHER

Here are some other invigorating drinks to keep you hydrated and healthy. Mix all ingredients in a jar, seal, and shake.

Super Energy Drink

- 1 cup (240 ml) warm water
- 1 tablespoon (20 g) unfiltered raw organic apple cider vinegar
- 3 tablespoons (60 g) raw honey
- 2 tablespoons (20 g) bee pollen (optional)

Orange Mint Medley

- 1 cup (240 ml) mint tea
- ½ cup (120 ml) coconut water
- ½ cup (120 ml) freshly squeezed orange juice
- 2 tablespoons (40 g) raw honey
- pinch of sea salt

Fig. 3: *Seal the jar and shake vigorously to blend.*

HONEY STRAWS

YOU WILL NEED

- plastic straws
- scissors
- raw unfiltered honey in a squeeze bottle
- pliers
- candle
- matches
- damp cloth

BEE BUZZ

One worker bee will produce about 1/12 teaspoon (0.5 g) of honey in her lifetime.

FUN FOR KIDS

Try sealing one end of the straw before filling with honey. What happens?

Make your own honey straws to carry with you for an anywhere, anytime instant energy boost.

DIRECTIONS

1. Any plastic straws will work, but clear or translucent straws allow you to see the amount of honey in the straw. If using flexible straws, cut off the bendable part of the straws before filling. Cut the straws to the desired length.

2. Connect the opening of the squeeze bottle to the straw. The respective sizes of the straw and the bottle's spout will dictate whether the straw will fit inside or outside of the spout's opening. If possible, use a straw that will fit outside the bottle's spout, as a straw inserted into the spout is more prone to drips.

3. Gently squeeze the honey into the straw, leaving at least 2" (5 cm) of space at the top. As much as possible, avoid stopping and starting the honey flow, as it will result in air bubbles. (Fig. 1)

4. Carefully remove the straw from the bottle's spout, keeping the straw level.

5. Pinch the clean end of the straw with pliers, exposing only a small amount of the plastic. (Fig. 2)

6. Tilt the straw so the honey flows toward the end with the pliers.

7. Light a candle. Carefully hold the pinched end of the straw in the flame until the plastic has melted. Remove the straw from the heat while continuing to clamp the pliers for a few seconds to complete the seal. (Fig. 3)

8. Stand up the straw, sealed-end down, to allow the honey to drain away from the unsealed end.

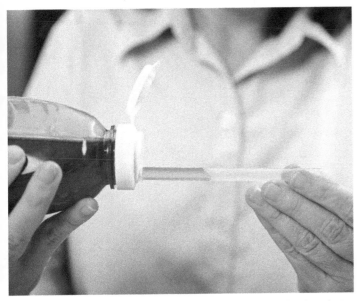

Fig. 1: *Using constant pressure, squeeze honey into a straw, leaving plenty of space at the end.*

Fig. 2: *Clamp pliers so only a small piece of the straw's end can be seen.*

9. The second seal can be a little tricky due to the honey residue in the straw. To remove as much remaining honey as possible before sealing, pinch the end using the pliers, wipe the seeping honey from the straw's opening with a damp cloth, and then hold it in the flame for a few seconds. If the seal isn't complete after a few seconds, it may be necessary to hold the straw's end in the flame a second time to complete the seal.

10. Rinse the finished honey straws in water before storing.

11. Slip a few honey straws into your backpack for a quick pick-me-up during a hike or day trip.

🐝 TAKE IT FURTHER

Using this technique, make a portable salve or cream pack by inserting propolis salve (Lab 34) or beeswax cream (Lab 36) into a straw using a syringe.

Fig. 3: *Using the pliers, hold the straw's end in a candle flame for a few seconds to seal the straw.*

LAB 13

HERBAL-INFUSED HONEY

YOU WILL NEED

- 1 orange
- grater
- 1 mint tea bag
- jar with tight-fitting lid
- 1 cup (340 g) raw, unfiltered honey (light, mild honey)
- strainer (optional)
- label and marker

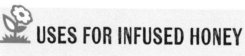

USES FOR INFUSED HONEY

- teas and smoothies
- salad dressings and sauces
- dessert recipes
- baked goods
- body-care products, such as creams and scrubs

Complement the flavor of your honey with herbs and spices. This particular blend uses two nectar-producing plants that bees love: mint and orange.

DIRECTIONS

1. Wash the orange before grating the peel to make orange zest. Avoid grating to the bitter white pith just beneath the surface. Measure out 1 tablespoon (6 g) of zest. (Fig. 1)

2. Place the mint tea bag in the jar along with the orange zest.

3. Pour the honey into the jar, completely covering the zest and tea bag. (Fig. 2)

4. Tightly screw on the lid. Place on a sunny windowsill and let steep for about a week, inverting the jar occasionally to keep the herbs coated in honey. (Fig. 3)

5. Remove the tea bag. Eat the orange zest along with the honey or strain to remove it.

6. Label the jar with the ingredients and the date it was made. Enjoy!

BEE BUZZ

There are at least 200 substances in most honeys, including sugars, enzymes, minerals, salts, acids, proteins, trace elements, amino acids, vitamins, fats, lipids, and more.

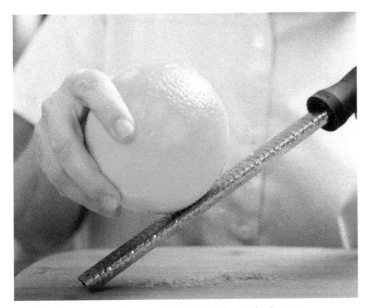

Fig. 1: *Make orange zest by grating the peel of a washed orange.*

Fig. 2: *Cover the mint tea bag and orange zest with honey.*

TAKE IT FURTHER

- Create your own herbal blends for infused honey. A good ratio is 1 to 2 tablespoons (3 to 6 g) of dried herbs to 1 cup (340 g) of honey. Possible flavors to choose from, or combine, include rosemary, lavender, chamomile, rose petals, vanilla beans, cinnamon, star anise, and cardamom.

- If using fresh herbs, begin by dividing the honey to be used in half. This is a great use for crystallized honey. Warm half of the honey and all the herbs in a double boiler over very low heat for 10 to 15 minutes. To make a double boiler, place a stainless steel saucepan in a larger pan of water. Remove the pan from the heat and let the herbs steep for at least 30 minutes or overnight and then remove the herbs. If using very strong flavorings, such as jalapeños or garlic, remove them immediately. Blend in the other half of the honey. This will help decrease the water content in the infused honey, thereby avoiding fermentation.

Fig. 3: *Infuse the honey and herbs for a week on a warm windowsill.*

LAB 14

HONEY-TASTING PARTY

YOU WILL NEED

- variety of honeys
- paper
- pens
- squeeze bottle for each honey
- labels
- spoons
- water to clean the palate

BEE BUZZ

In the United States alone, there are more than 300 types of honey from different floral sources.

FUN FOR KIDS

How does the taste of the darkest colored honey compare to that of the lightest honey? Does the particular season of the honey (summer or fall) make a difference in color or taste?

Celebrate the amazing partnership between flowers and honeybees by holding a honey-tasting party. Guests try to match the honey tasted with the bees' nectar source.

DIRECTIONS

1. First, identify and select a few unique artisan honeys to taste. There are a number of ways to acquire a variety of honeys:

- Collect honey from your hives at different times of the honey season.

- Purchase at a farmers' market, specialty store, or health food store.

- Ask beekeeping friends to bring their honey to the party.

- Trade honey with other beekeepers.

- Obtain honey when you travel.

2. Create a numbered list of each honey to be tasted. Write a brief description of each honey, including the season, year, location, possible nectar sources, and any other known information.

3. Pour each honey into a separate squeeze bottle, and then label each bottle with a different color or letter. (Fig. 1)

4. You may also consider pouring each honey into small, labeled glass containers to better compare the honey colors.

5. Create the answer key for the matching game by marking one copy of the list with the bottle colors or letters that correspond to each honey description number.

6. Give each guest a spoon and a copy of the honey description list.

7. Have each guest squeeze a little honey onto a spoon, smell and taste the honey, and then jot down notes about the flavor and color. Guests will continue tasting each honey, cleansing their palates with sips of water between samples. (Fig. 2)

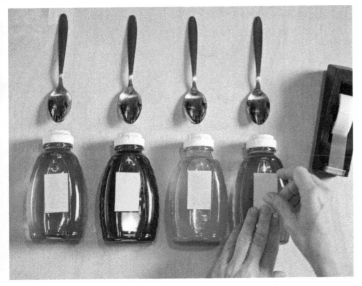

Fig. 1: *Label each honey to be sampled with a different color.*

Fig. 2: *Squeeze onto a spoon to taste each honey.*

8. Invite guests to guess the different honeys by writing down the colors or letters on the honey bottles next to the matching honey descriptions on the list. (Fig. 3)

9. Announce the names and descriptions for each bottled honey. Give prizes or awards to the people with the most correct matches.

10. Have guests cast votes for the favorite honey to receive a "People's Choice" award.

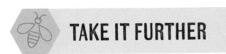

TAKE IT FURTHER

- Have a honey-tasting party using herbal-infused honeys (Lab 13).

- Take a culinary challenge by partnering one of these artisan honeys with a specific food choice to complement both flavors.

Honey Number	Description	Assigned Color
1	fall honey in an area full of goldenrod	
2	wildflower honey from the area	
3	tupelo honey from Florida	

Fig. 3: *Here is a sample of what the guests may be using to match the honey tasted with the descriptions given.*

UNIT 3

BEESWAX

THE INDUSTRY AND DILIGENCE HONEYBEES NEED FOR COMB BUILDING
PROVIDES A FOUNDATION FOR THE HIVE AND AN EXAMPLE FOR LIFE.

Wax production is one of the first steps in building a new bee home.
Honeybees expend an incredible amount of energy to gather and convert nectar
into honey, to consume the honey necessary for wax production, and to build the comb
essential for brood rearing and food storage. I do not want to take the efforts
of the bees for granted, so I save and use every little bit of beeswax.

Beeswax, in and of itself, offers myriad uses. It can be used to make candles,
soap, artwork, skin care products, batik prints, mustache wax,
furniture polish, and leather conditioners, and is used in dental procedures,
to help start fires, for waterproofing, and for metal casting.

The activities in this unit serve as a jumping-off point for making
beeswax products. Each time you light a beeswax candle, wear jewelry made
with wax, sculpt with waxed thread, or use wax for utilitarian purposes, thank the
honeybee for its industry. Every little bit of beeswax is precious.

Design. Make. Create. Enjoy.

ROLLED BEESWAX CANDLES

YOU WILL NEED

- 1 beeswax sheet, also called foundation, 8" × 16" (20.5 × 40.5 cm)
- 2/0 wick
- scissors
- hair dryer (optional)

BEE BUZZ

Bees build their comb with the openings of the cells slanting upward by 9 to 13 degrees.

FUN FOR KIDS

Cut the wax sheets into fourths for smaller hands to roll the candles with ease.

Rolling beeswax foundation, or wax sheets stamped with the base of a cell, is a simple, safe, and inexpensive way to make lovely candles without the need for hot melted wax.

DIRECTIONS

1. Fold the beeswax sheet in half with the short ends together. Bend the sheet back and forth until it breaks. Each piece will make one candle. (Fig. 1)

2. Lay the wick on the torn edge of the wax. Cut the sheet a little longer than the end of the sheet.

3. Use your fingertips to crimp and press the wax around the wick. (Fig. 2)

4. Tightly roll the wax using both hands to exert even pressure. If the beeswax sheet is cold, blow hot air over the wax for a few seconds with a hair dryer before rolling. (Fig. 3)

5. Use the warmth of your fingertips to gently press the wax seam closed. (Fig. 4)

6. Burn these candles or bundle them together with colorful ribbon to give as a beautiful gift.

Fig. 1: *Fold and break a beeswax sheet in half.*

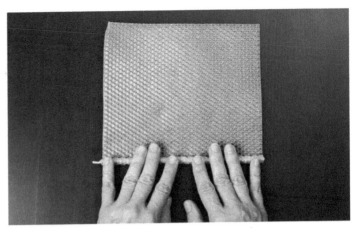

Fig. 2: *Press the wax around the wick.*

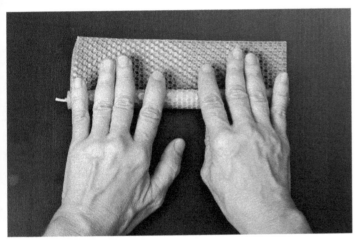

Fig. 3: *Evenly roll the wax with both hands.*

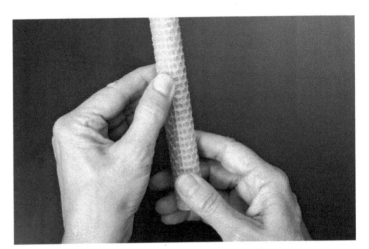

Fig. 4: *Seal the seam using slight pressure from your fingertips.*

 ## TAKE IT FURTHER

- Cut the beeswax in a variety of lengths and sizes to create candles of different shapes and diameters.

- A pizza slicer makes a great beeswax-cutting tool.

- Decorate the outside of a rolled candle by cutting colored wax sheets into shapes and strips. Press onto the surface of the candle with your fingertips.

- Try rolling two different colored triangular sheets together and slightly offset. When rolling, make sure there is one end side to fit in a candleholder.

HAND-DIPPED CANDLES

YOU WILL NEED

- newsprint or cardboard
- double boiler
- beeswax
- large coffee can or tall metal container
- 2/0 or 1/0 square braid wicking
- hex nuts
- scissors or sharp knife
- tall container of cool water
- candy or deep-frying thermometer
- ruler and bucket

BEE BUZZ

Because honeybees need to gather and consume nectar and/or honey to produce wax, it takes bees seven times more energy to make the wax than it does to produce honey.

SAFETY NOTE

Keep wax away from open flames. Unplug heating source when not in use.

Dipping candles has a way of taking us back to another time and place. All you really need is some wax and a wick, and time to make these useful candles.

DIRECTIONS

1. Set up a workstation with newsprint to protect your work surface and cardboard to protect the floor.

2. In a double boiler, melt enough beeswax to fill a tall metal container about 1" (2.5 cm) from the top.

3. To dip two candles at once, cut the wick about two and a half times the length of the dipping container. Tie a hex nut at each end of the wick. (Fig. 1)

4. Hold the wick in the middle. Lower into the wax to soak for a few seconds. (Fig. 2)

5. If the wick is not smooth when removed from the wax, cool it a bit and then run your fingers down each side.

6. Begin the dipping process. Dip in the wax and then dip in cool water. (Fig. 3)

7. Pull down on the hex nuts occasionally to straighten the wicks, especially toward the beginning of the process. Dip until the desired diameter is reached.

8. Cool a bit before cutting off the hex nuts with scissors or a sharp knife. Remelt the wax end. (Fig. 4)

9. Increase the wax temperature to 180°F (82°C) (check with a thermometer) and dip one last time.

10. Hang the candles to dry. A simple hanger for shorter candles can be made by laying a ruler over the top of a bucket or pot. (Fig. 5)

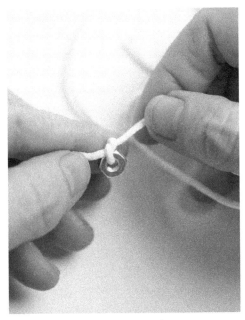

Fig. 1: *Weight the wicks by tying hex nuts to each end.*

Fig. 2: *Dip the wick in the melted wax.*

Fig. 3: *Dip the wick in cool water and then continue alternating between the wax and water until satisfied with the diameter of the candle.*

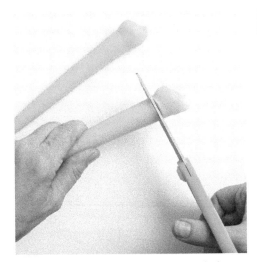

Fig. 4: *Cut off the wax-covered hex nuts before finishing with one last wax dip.*

Fig. 5: *Hang undisturbed until completely dry.*

TAKE IT FURTHER

Try flattening a warm dipped candle with a rolling pin. Holding the top and bottom, twist the candle into a spiral. Dip one last time.

FUN FOR KIDS

What other materials besides beeswax can be used to make candles?

TEA LIGHTS

YOU WILL NEED

- newsprint or cardboard
- beeswax
- double boiler
- candle release
- uncoated minimuffin pan
- 2/0 square braid wicking
- scissors
- metal wick clips
- pliers
- candy or deep-frying thermometer

These small candles are very forgiving when it comes to finding the correct wicking. Float these lovely candles in a bowl of water or place in a votive holder.

DIRECTIONS

1. Set up a workstation with newsprint or cardboard to protect your work surface.

2. Melt the wax in a double boiler.

3. Spray candle release on muffin tins that are not coated with a nonstick surface.

4. Dip a 2' (60 cm) piece of wicking into the melted wax. (Fig. 1)

5. Cut the wax-coated wick into 1½" (3.8 cm) pieces. Feed one end of the wick into a wick clip. Crimp closed with pliers. (Fig. 2)

6. Place a wick in each cup of the muffin pan.

7. Take the wax off the heat source to cool down for a few minutes. Check the temperature with a thermometer; it should be 150°F to 160°F (65.5°C to 71°C). Pour the wax into each cup. (Fig. 3)

8. Allow the wax to cool completely. Pop out the tea lights by turning over the pan and with a little force knock it with your hand or on the table. If you are having a hard time getting them out, freeze the tray and then try again.

9. Clip the wicks. These candles will burn longer if placed in a votive holder.

BEE BUZZ

A healthy hive with about 50,000 bees should be able to produce ½ pound (227 g) of wax a day.

SAFETY NOTE

Keep wax away from open flames. Unplug heating source when not in use.

Fig. 1: *Coat a long piece of wick in melted wax.*

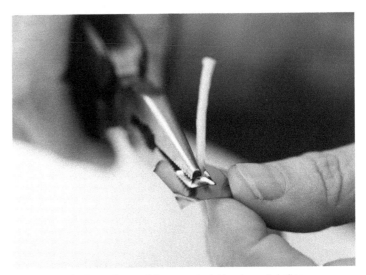

Fig. 2: *Attach short pieces of the waxed wicking to metal clips using pliers.*

![bee icon] **TAKE IT FURTHER**

- Use these tea lights with the beeswax candle bowls (Lab 18).

- Try your hand at making regular votive candles. Use small paper cups as molds or purchase store-bought metal molds. You will probably need a larger sized wick (start with #2).

Fig. 3: *After the wax has cooled a bit, carefully pour into each cup.*

BEESWAX CANDLE BOWLS

YOU WILL NEED

- newsprint or cardboard
- electric hot plate
- double boiler (the top part must be at least 7" [17.8 cm] in diameter)
- beeswax
- water balloons
- paper towels
- glue stick
- pressed flowers
- paring knife
- griddle or warming tray
- aluminum foil
- tape
- ladle
- sand
- tea light

⚠ SAFETY NOTE

To avoid dropping the balloon and splashing hot wax, children may need help holding the balloon while dipping.

Grouped together, candle bowls make an impressive presentation.

One tea light can transform this delicate beeswax candle bowl votive into a warm, luminous lamp. Impress your family and friends with this easy-to-make gift using nothing more than a wax-covered water balloon and dried flowers.

DIRECTIONS

1. Set up a workstation with newsprint or cardboard to protect your work surface. Plug in the hot plate and put it on your work surface.

2. Place the double boiler on the hot plate and melt the wax. You will need about 6" (15 cm) of molten beeswax. Leave plenty of space between the top of the wax and the top of the container for wax displacement.

3. Make a water balloon to dip in the wax by stretching the mouth of the balloon over a faucet. Slowly run the water while firmly supporting the bottom of the balloon as it expands. Tightly squeeze the mouth of the balloon while removing it from the faucet. Tie a knot at the top of the balloon. Dry the water balloon completely with a paper towel.

4. Using a smooth, fluid movement, dip the water balloon in and out of the wax slightly past the balloon's halfway point. Do not stop in the middle of the dipping movement or it will result in a visible seam. Wait a few seconds and then dip the water balloon into the wax again to the same depth. Repeat approximately 20 times to create a durable thickness of wax. The hotter the wax, the thinner each coat will be, so additional dips may be needed. (Fig. 1)

5. Cradle the wax-coated water balloon in your lap or on a towel. Use a glue stick to attach the dried flowers and leaves onto the wax. (Fig. 2)

6. Dip the water balloon into the hot wax one last time to coat and seal the flowers. Set the water balloon upright to cool for a few minutes.

Fig. 1: *Dip the balloon in the wax using one fluid motion.*

Fig. 2: *Decorate the balloon by gluing pressed flowers on the outside.*

Fig. 3: *Smooth the rim and flatten the base of the candle bowl using very low heat.*

Fig. 4: *Add beeswax to strengthen the base of the candle bowl.*

Fig. 5: *Add sand to insulate the candle bowl from the tea light's heat.*

7. Carefully puncture the water balloon over a sink using a small paring knife. The punctured water balloon will pull away from the wax sides, creating the candle bowl.

8. Cover the griddle or warming tray with aluminum foil and secure with tape. Turn the griddle to the very lowest setting possible. Smooth the rim of the candle bowl by placing the rim on the griddle. Turn the bowl right side up. Place it on the griddle, make sure it is level, rest your palm on the rim, and gently press down for a few seconds to make a flat base. Be careful not to completely melt the bottom. (Fig. 3)

9. Allow the melted wax to cool. Once cooled, using a ladle, carefully spoon a little melted wax into the candle bowl to strengthen the base. (Fig. 4)

10. Put sand in the bowl to insulate the bottom from the heat of a tea light. (Fig. 5)

 TAKE IT FURTHER

- Pressing flowers is great fun in itself. To allow for adequate reseeding, only pick flowers where there are at least ten plants present. Pick fewer than a third of the flowers in any one area. Make a simple plant press using recycled paper sandwiched between corrugated cardboard and held together with rubber bands.

- Not all flowers and leaves maintain their colors when pressed. Pansies, verbena, and larkspur maintain their colors well. Ferns, fennel, and dill add a beautiful feathery look.

- Challenge yourself by using only flowers that are nectar and pollen producers for honeybees.

BEESWAX JEWELRY

YOU WILL NEED

- pen
- plastic bottle cap
- white cardboard
- scissors
- beeswax sheet (also called beeswax foundation)
- hair dryer (optional)
- clear-drying glue
- coins for weights
- small paintbrush
- hammer
- small nail
- block of wood
- 2 metal jump rings, ½" (12 mm)
- 2 pairs of pliers
- 2 charms (optional)
- 2 ear wires

BEE BUZZ

The cell walls of the comb in a hive are 0.0003" (0.007 mm) thick.

Jewelry made with beeswax sheets is lightweight, inexpensive, and unique. Before you begin, think about the size and shape of the piece of jewelry you would like to create. This particular project uses circles of beeswax sheet to make a pair of earrings.

DIRECTIONS

1. To make a template, trace the bottle cap onto the cardboard and cut out.

2. Hold the cardboard circle firmly against the beeswax sheet and carefully cut around the perimeter. In cold weather, warm the beeswax sheet briefly with a hair dryer to eliminate cracking. Repeat this process to create a total of four wax circles. Melt the wax scraps for other projects. (Fig. 1)

3. Cut a thin strip from around the perimeter of each cardboard circle, making them slightly smaller. The reduced size will hide the cardboard from view in the finished product. (Fig. 2)

4. For each earring, sandwich a trimmed cardboard circle between two beeswax circles. Glue in place. (Fig. 3)

5. Improve the contact between the wax and cardboard by weighing it down with a few coins. Allow the beeswax/cardboard circles to dry completely.

6. After the beeswax/cardboard circles have dried, seal the edges by gently pinching the perimeter of each earring with your fingers. (Fig. 4)

7. Seal each beeswax/cardboard circle with glue by dipping a paintbrush in glue and dabbing it on the indented cells to coat. When the first side is dry, turn it over and repeat on the other side. Allow to dry completely overnight.

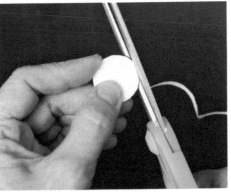

Fig. 2: *Trim carefully around the outer edge of each cardboard circle to reduce its size.*

Fig. 3: *Glue the smaller cardboard circle between two beeswax circles.*

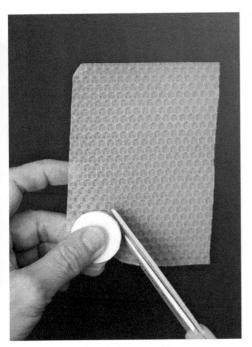

Fig. 1: *Cut four identical pieces of beeswax sheet using a cardboard circle as a template.*

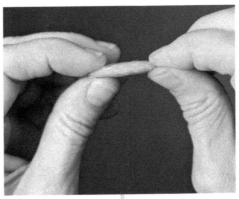

Fig. 4: *Pinch the perimeter of the beeswax circles with your fingertips to seal.*

Fig. 5: *Make holes to hang the earrings using a hammer and nail.*

8. Punch a hole near the edge of each beeswax/cardboard circle using a hammer, a small nail, and a block of wood. (Fig. 5)

9. Open a jump ring by holding each side of the split with pliers and twisting in opposite directions. Insert the jump ring into the hole of a beeswax/cardboard circle. If desired, slide a charm onto the jump ring. Attach ear wires to the jump ring and close with pliers, twisting the sides back toward each other. Repeat the process for the second earring.

10. Now you are ready to accessorize—thanks, in part, to the bees!

FUN FOR KIDS

Small hands need larger pieces to work with. Make a pendant by gluing pieces of beeswax foundation directly onto white cardboard or colored mat board. Create a hole using a hole punch instead of a hammer and nail. Hang the pendant using colored yarn.

TAKE IT FURTHER

• How can you incorporate bee pollen pellets into a piece of jewelry?

• Create a mini collage by layering bits of colored beeswax sheets on the base shape.

LAB 20

WAXED-THREAD ORNAMENTS

YOU WILL NEED

- newsprint or cardboard
- beeswax
- double boiler or slow cooker
- scissors
- embroidery thread
- wax paper
- thick marker

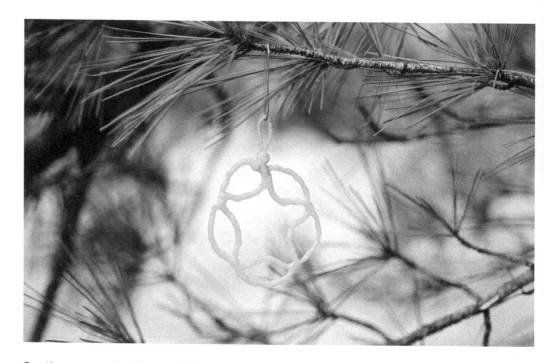

Creating ornaments with waxed thread will keep the entire family engaged. The design possibilities are endless. No need for glue, tape, or staples—the wax sticks to itself.

DIRECTIONS

1. Set up a workstation with newsprint or cardboard to protect your work surface. Melt the beeswax in a double boiler or slow cooker.

2. Cut a skein of embroidery thread into pieces about 2′ (60 cm) long.

3. Dip each strand of thread into the melted beeswax. Lay each waxed strand flat on wax paper to allow it to harden. (Fig. 1)

4. Twist two waxed strands together by twirling them between your fingers on one hand while holding the strand stationary with the other. The stickiness of the wax will bind the twisted strands together. (Fig. 2)

Fig. 1: *Dip each strand of embroidery thread into melted wax.*

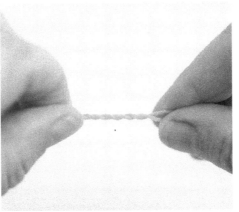

Fig. 2: *Twist two waxed strands of embroidery thread together.*

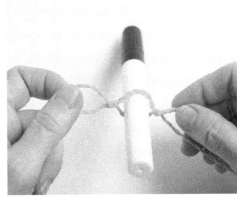

Fig. 3: *Make five ovals by using a thick marker as a spacer. Twist the strand together to hold in place.*

5. You will need five evenly sized loops to make this particular waxed thread ornament. Bend a twisted strand in half. Make the first loop by placing the marker in the fold. Twist the strands together with a half turn around the marker. Continue making a total of five loops along the strand using the marker as a spacer. (Fig. 3)

6. Bend the strand into a wreath shape. Attach the two ends by threading the loose end of thread through the end of the loop. (Fig. 4)

7. Make a ring to hang the ornament by looping one thread end at the top and then wrapping with another piece of thread at the base to secure. Carefully clip all thread ends. (Fig. 5)

8. Coat the ornament by dipping it into the melted wax as many times as desired.

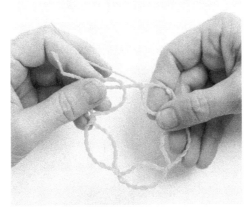

Fig. 4: *Thread the end strands through the last loop to complete the circle.*

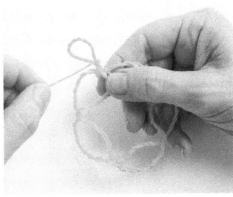

Fig. 5: *Bend and wrap the strands around the top to make a loop for hanging the ornament.*

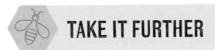

TAKE IT FURTHER

What other shapes and patterns can you create using waxed thread?

FUN FOR KIDS

Play with a few different colored waxed threads to create a two- or three-dimensional sculpture.

WATERPROOF BAG

YOU WILL NEED

- newsprint or cardboard
- beeswax
- soy wax
- double boiler or slow cooker
- small canvas bag
- paintbrush
- hair dryer
- paper towels

BEE BUZZ

Beeswax is made up of more than 300 individual components.

⚠ SAFETY NOTE

Keep wax away from open flames. Unplug heating source when not in use.

Wax-impregnated fabric repels water beautifully. Create an all-natural alternative to petroleum products by using a combination of beeswax and soy wax. The bag for this project is a canvas lunch tote.

Fig. 1: *Completely cover the bag using a brush and mixture of melted beeswax and soy wax.*

DIRECTIONS

1. Set up a workstation with newsprint or cardboard to protect your work surface.

Melt approximately one part beeswax and one part soy wax in a double boiler made by placing a can in a pan of water. Using 100 percent beeswax is not advisable because it is difficult for the canvas to fully absorb. The beeswax provides a natural yellow color.

2. Coat the bag by brushing a thin layer of the hot wax mixture onto the canvas. Make sure the entire outside of the bag is covered, including the seams. As the wax dries, turn the bag to coat all sides, including the bottom. Let the bag thoroughly harden. (Fig. 1)

3. Heat is needed to embed the wax in the canvas. Use a hair dryer on the highest setting for this process. While holding paper towels inside the bag to absorb any excess wax, slowly move the hair dryer to blow hot air across the surface of the bag, allowing the wax to be absorbed into the fabric. (Fig. 2)

4. Enjoy using your waterproof bag!

Fig. 2: *A hair dryer provides a heat source to embed the wax in the canvas.*

 TAKE IT FURTHER

Waterproof other canvas bags used for cameras, cosmetics, camping gear, or backpacks.

FUN FOR KIDS

Before waterproofing, make the bag your own by adding artwork using fabric paints or comb rubbings (Lab 44).

LEATHER CONDITIONER

YOU WILL NEED

- newsprint or cardboard
- beeswax
- coconut butter
- double boiler
- almond oil
- lemon essential oil (optional)
- wooden stick
- small containers

BEE BUZZ

Bees warm the wax to about 109°F (43°C) while constructing the hexagonal cells.

⚠ SAFETY NOTE

Keep wax away from open flames. Unplug heating source when not in use.

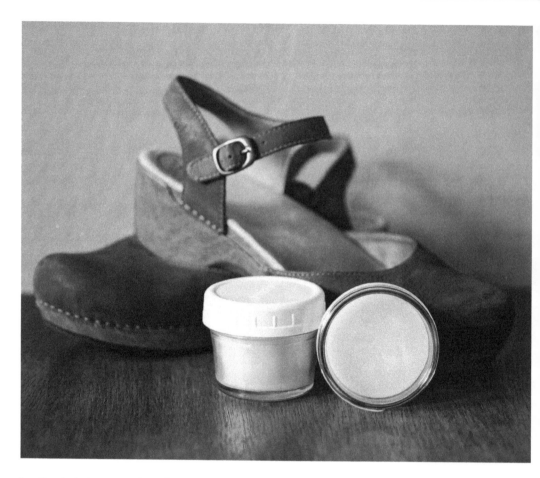

Leather lasts longer and retains its natural beauty when conditioned on a regular basis. Applied to leather, beeswax protects, coconut butter conditions, almond oil softens, and lemon essential oil prevents cracking.

Fig. 1: *Add the lemon essential oil.*

Fig. 2: *Pour into a container with a lid.*

DIRECTIONS

1. Set up a workstation with newsprint or cardboard to protect your work surface. Melt one part each beeswax and coconut butter in a double boiler. A tin can in a pot of water will do the job.

2. Add two parts almond oil. Once everything has melted, remove the can from the double boiler.

3. Allow the mixture to cool down for a few minutes before adding 15 drops of lemon essential oil for every ½ cup (120 ml) of conditioner. Stir with a wooden stick. (Fig. 1)

4. Pour into a small container with a lid. (Fig. 2)

5. To use, rub the conditioner into leather using your fingertips. Once finished, buff the leather using a clean, dry cloth. You may want to test this conditioner on a small hidden area before using it on the entire item.

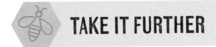 **TAKE IT FURTHER**

For thicker conditioner, increase the amount of beeswax and coconut butter. If you are looking for something more along the lines of a cream, add more almond oil.

FIRE STARTERS

YOU WILL NEED

- newsprint or cardboard
- beeswax
- double boiler or slow cooker
- newspaper
- 1" (2.5 cm)-thick marker
- masking tape
- tweezers

BEE BUZZ

Beeswax has a melting point of 145°F (64°C).

SAFETY NOTE

Use only with adult supervision. Unplug heating source when not in use.

Beeswax is very flammable. That is a good thing if you are camping and want to build a fire. These simple compact fire starters are easy to make, waterproof, and lightweight—perfect for an outdoor adventure.

DIRECTIONS

1. Set up a workstation with newsprint or cardboard to protect your work surface. Melt the beeswax in a double boiler by placing a can in a pan of water or use a slow cooker.

2. Tear newspaper into strips 3" to 4" (7.5 to 10 cm) wide by 22" (56 cm) long. The measurements are flexible. (Fig. 1)

3. Roll a newspaper strip around your fingers or a thick marker to make a tube. The inside diameter of the tube should be about 1" (2.5 cm) wide. Tape the end to secure. (Fig. 2)

4. Using tweezers, grab one end of a newspaper tube. Dip the tube into the melted wax to completely coat it. If the wax is not deep enough to fully immerse the tube, dip one end of the tube, allow it to cool, and then dip the other end. (Fig. 3)

5. Cool the newspaper tube on a piece of paper. Store the fire starters in a paper or plastic bag.

6. Build a nice warm fire with ease on your next camping trip.

Fig. 1: *Tear newspaper into long, thin strips.*

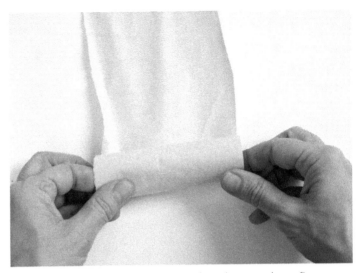

Fig. 2: *Make newspaper tubes by rolling the strips around your fingers or a marker.*

Fig. 3: *Dip the paper tubes into the melted beeswax.*

 ## TAKE IT FURTHER

- Use these fire starters in your smoker as long as they are made with 100 percent beeswax and no paraffin.

- Experiment to find the longest burning time for these fire starters by changing one variable at a time: length, width, thickness, and kind of paper.

FUN FOR KIDS

Scouting groups, under close adult supervision, will enjoy making and using these waterproof campfire starters. These fire starters almost guarantee a one-match campfire.

UNIT 4

POLLINATION

HONEYBEES REMIND US THAT ALL THINGS ARE INTERRELATED.

Pollination makes the world go round. Pollinators make it possible for many flowering plants to produce seeds. In turn, flowering plants provide food for the pollinators. The reduction or loss of either the pollinator or the pollinated will ultimately affect the survival of both. Humans, wildlife, and flowering plants depend on the life cycle pollination provides.

The value of insect pollination is immense and difficult to measure. We derive nutritional benefits from fruits, nuts, seeds, and vegetables; health benefits from plant-based medicines; and economic benefits from agricultural production.

Although pollination's impact is global and far reaching, I find a simple joy in watching bees fly from one flower to the next. It is nearly impossible for me to pass a plant buzzing with bees and not stop to take it all in—the bees collecting food, the creation of seeds, and the delicate dance of pollination.

Take a closer look and experience the flower and bee connection of pollination through the experiments and activities in this unit.

Observe. Examine. Explore. Pollinate.

POLLINATION MAGIC

YOU WILL NEED

- 2 sheets of black construction paper
- glue
- scissors
- paper clips
- 4" × 6" (10 × 15 cm) photo or drawing of an apple
- 4" × 6" (10 × 15 cm) photo or drawing of an apple blossom
- 4" × 6" (10 × 15 cm) photo or drawing of a honeybee

Amaze your family and friends with this simple magic trick using a hidden pocket. Change an apple blossom into an apple right before their very eyes! This is a wonderful attention getter for any presentation about honeybees.

DIRECTIONS

1. Fold one sheet of black paper so one side is about a finger's width longer than the other side. (Fig. 1)

2. To make the hidden pocket, open the paper. Place a line of glue along the fold and the top and bottom edges of the shorter side. (Fig. 2)

3. Lay the other piece of black paper on the glue with one end against the fold.

4. Fold the paper in half to make a folder. Cut off the glued sheet of paper to the exact length of the longer side. This will make a pocket in the back of the paper folder. (Fig. 3)

5. Set up the pollination trick beforehand by placing the photo or drawing of the apple in the paper folder.

6. Use paper clips to fasten the sides of the folder closed.

7. It's show time! Hold the folder so the hidden pocket is toward you. Hold up the image of the apple blossom and say something like, "We are going to change this apple blossom into an apple thanks to pollination." Slip the apple blossom into the hidden pocket of the magic-change envelope. To the audience, it looks like the picture is going directly into the folder. Ask the audience what that blossom needs to become an apple. Once someone suggests a pollinator, pull out the image of the honeybee.

BEE BUZZ

Who pollinates crops? Here is the breakdown: 72.7 percent bees, 18.8 percent flies, 6.5 percent bats, 5.2 percent wasps, 5.1 percent beetles, 4.4 percent butterflies and moths, 4.1 percent birds, and 1.3 percent thrips.

FUN FOR KIDS

To help children make the connection between flowers, bees, and pollination, make a stick puppet with a picture of an apple on one side and a flower on the other. Use your finger as a bee to pollinate the flower. Spin the stick puppet around to transform into an apple.

Fig. 1: *Fold a sheet of black paper so one side is a little longer than the other.*

Fig. 2: *Place glue along the fold and the shorter edges of the shorter side.*

Fig. 3: *Fold and cut off the glued paper to match the length of the longer side.*

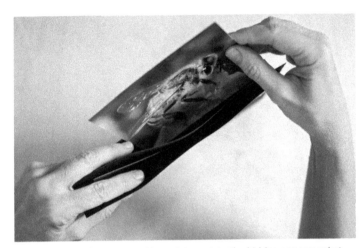

Fig. 4: *Slide the blossom and bee pictures into the hidden paper pocket.*

Slip the honeybee into the hidden pocket. If working with children, have everyone throw their pretend bees into the envelope to make sure the blossom is fully pollinated. (Fig. 4)

8. Conjure up some magic words. Here is an example of the words I use. "Pollination, pollination, pollination. Abracadabra zimmity zam. Turn into an apple, if you can!" With drama and flair, flip open the folder to expose the apple image. Magically the blossom and honeybee have disappeared!

TAKE IT FURTHER

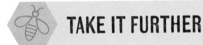

- Can you think of other ways to use the magic-change envelope?

- Instead of paper, make a magic-change bag using fabric and a fastener such as Velcro.

LAB 25

DISSECT A FLOWER

YOU WILL NEED

- lily
- magnifying glass
- craft knife
- cutting board
- paper
- tape
- pen

Fig. 4: *Reconstruct the flower on a piece of paper.*

Examine a lily to learn about the reproductive parts of a plant that are essential for pollination.

DIRECTIONS

1. Pick a lily from your yard, buy one from a store, or visit a local florist to obtain discarded flowers.

2. Remove the sepals and petals by cutting or pulling them off the stem. In a lily, the bottom three petal-like structures are the sepals. (Fig. 1)

3. Examine the male parts of the flower. Detach the stamens. Identify the anthers, pollen, and filaments. (Fig. 2)

4. Inspect the pistil, the female parts of the flower. Look closely at the stigma using a magnifying glass. Separate the style from the ovary by cutting it with a craft knife on a cutting board. If possible, slice the style in half lengthwise. Observe the long hollow pollen tube of the style. (Fig. 3)

5. Carefully cut open the flower's ovary to expose the tiny eggs, or ovules, inside. Unfertilized eggs can be difficult to see.

6. Construct a visual record of the flower parts by laying each individual segment on a piece of paper. Tape it onto the paper and label it, if desired. (Fig. 4)

BEE BUZZ

Complete flowers contain sepals, petals, stamens, and pistil. Perfect flowers have both male and female parts.

FUN FOR KIDS

Encourage young children to identify the stem, leaves, and petals of a flower.

TAKE IT FURTHER

- Dissect composite flowers such as dandelions, sunflowers, zinnias, and daisies. Examine the sterile petal-like ray flowers and the center disk flowers that develop into seeds.

- Each seed in a sunflower is produced by a single disk flower. Count the clockwise and counter-clockwise spirals on the seed head and then count the petal-like ray flowers. Do your numbers correspond to the mathematical sequence called the Fibonacci sequence: 0, 1, 1, 2, 3, 5, 8, 13, 21, 34, 55, 89 . . . ?

Fig. 1: *Pull the sepals and petals off the flower's stem.*

Fig. 2: *Take a look at the male parts of a flower by removing the stamens.*

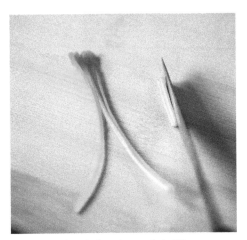

Fig. 3: *Cut the pistil and ovary in half for closer examination.*

 # PARTS OF A FLOWER

STAMEN: MALE FLOWER PARTS

Anther: produces and contains pollen grains

Filament: stalk that supports the anther

PISTIL: FEMALE FLOWER PARTS

Stigma: the sticky surface where pollen lands and germinates

Style: a narrow tube that carries germinated pollen from the stigma to the ovules (eggs) in the ovary

Ovary: contains ovules that become seeds once fertilized through pollination

OTHER PARTS

Sepals: small leaves that closely resemble petals and are directly under the flower to protect the bud prior to opening

Petals: brightly colored, modified leaves of a flower that attract pollinators

Stem: the stalk that supports the plant

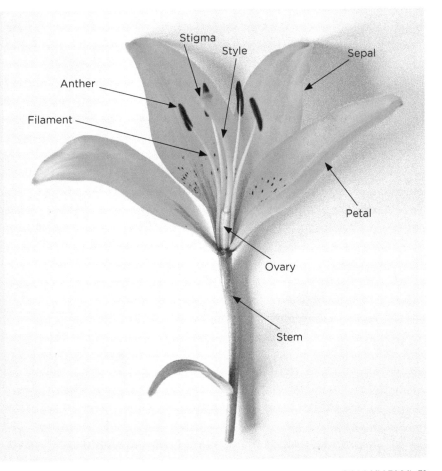

BEE LANDING STRIPS

YOU WILL NEED

- flowers
- ultraviolet (UV) flashlight
- thick blanket

BEE BUZZ

Bees can see yellow, blue-green, blue, violet, ultraviolet, and a color known as "bee's purple," a mixture of yellow and ultraviolet.

FUN FOR KIDS

Cut out photos of flowers from magazines and old calendars. Go into a dark space or cover up with a blanket. Use a UV flashlight to view any nectar guides that may be present.

Some flowers have developed an adaptation that benefits both flowers and bees—ultraviolet nectar guides that attract both bees and butterflies. This adaptation increases flowers' chances of reproduction through pollination and allows bees to find food more efficiently by rapidly locating nectar sources. Using ultraviolet light we can view the floral "landing strips" that are visible to bees but not to humans.

DIRECTIONS

1. First you will need a variety of flowers to view and compare nectar guides. Either venture outside and pluck a few samples or purchase a bouquet containing an assortment of flowers. Often a local florist will give away flowers past their prime for specific projects like this one.

2. Look at each flower under normal (full-spectrum) light. (Fig. 1)

3. Take a UV flashlight and the flowers into a dark space or create a lightless location under a heavy blanket. Shine the UV flashlight on each flower to observe the nectar guides, if present. (Fig. 2)

Fig. 1: *Gaillardia viewed under full-spectrum light.*

Fig. 2: *Gaillardia viewed under UV light.*

- What other markings and patterns do flowers have to attract pollinators?

- Describe the shape and color of flowers with nectar guides. Are there any similarities? Differences? Identify each flower using field guides or by consulting an extension office.

- Instead of picking flowers, view nectar guides outside. Kneel near the flowers. Block out as much sunlight as possible by covering both you and the plants with a thick blanket. View with the UV flashlight.

PORTABLE VIEWING STATION

Make a portable viewing station for nectar guides, perfect for use at outdoor events. Lay an opaque plastic bin or cardboard box on its side with the opening facing toward the front. Tape a piece of dark-colored fabric to the top edge of the bin. Place the flowers inside the open space of the bin. To use, position your head near the bin opening, and then flip the fabric over your head and back to create a dark viewing space.

CHARGED POLLINATION

YOU WILL NEED

- photograph of a flower
- paper plate
- black pepper
- round balloon
- piece of wool (optional)
- black permanent marker (optional)
- tissue paper (optional)

BEE BUZZ

After a bumblebee visit, a flower can become positively charged, letting other bees know that a pollinator has already been there. In this way, bees can efficiently move on to another flower loaded with nectar.

Bees bump into dust and small particles while flying. The friction removes electrons from the surface of the bee's body, creating a positively charged bee. Flowers tend to have a negative charge. Since opposite charges attract, the pollen can "jump" from the negatively charged flower to the positively charged bee. Illustrate this process by using a balloon, black pepper, and static electricity.

Fig. 1: *Sprinkle pepper on a paper plate flower to represent pollen.*

Fig. 2: *Build up static electricity by rubbing the balloon with a piece of wool.*

DIRECTIONS

1. Place a photograph of a flower on a paper plate. Shake pepper into the middle of the flower to represent pollen. (Fig. 1)

2. Blow up a balloon and tie it closed.

3. If desired, draw black stripes on the balloon using a permanent marker to signify a honeybee or bumblebee. Add tissue paper wings, if desired.

4. Imitate a bee flying through the air by rubbing the balloon on your hair, the carpet, or a piece of wool to build up static electricity. (Fig. 2)

5. Hold the charged balloon close to the paper plate flower without actually touching it. (Fig. 3)

6. What happens to the "pollen"?

Fig. 3: *What happens when the balloon nears the "pollen"?*

TAKE IT FURTHER

• Use the slow-motion video recorder on a smart phone to record the "pollen" moving from the "flower" to the "bee."

• Are there other objects, beside pepper, that the charged balloon will attract?

APPLE POLLINATION

YOU WILL NEED

- apples
- sharp knife
- cutting board
- butter knife

Successful apple and seed production depends on cross-pollination, or the transfer of pollen between two different varieties of apple trees. This simple experiment reveals the effectiveness of pollination in one apple blossom.

DIRECTIONS

1. Observe an apple, noting the shape, color, and general appearance.

2. Using a sharp knife on a cutting board, cut the apple in half around the middle like the equator of the earth. (Fig. 1)

3. Examine the five carpels, or seed pockets, on each half of the apple. The pistil of the apple blossom is made up of five carpels. Each carpel segment has a stigma, a style, and a portion of the ovary. They are pollinated separately.

4. Using a butter knife, scrape the seeds out onto the fruit's surface. (Fig. 2)

5. Count each fully developed, viable seed. A fully pollinated apple will have ten seeds.

Note: Some possible causes for incomplete pollination include reduced pollinator visits, lack of apple tree varieties for cross-pollination, damaged flowers from a freeze or storm, low viability of pollen, or declining tree health.

BEE BUZZ

It takes two or three hives, or 120,000 to 180,000 bees, to pollinate 1 acre (0.4 ha) of apple trees.

FUN FOR KIDS

Cut open ten apples. Mix in math by making a three-columned chart: description of the apple, number of fully developed seeds, and number of shriveled seeds. Which apples had the most and least pollinator visits? Calculate the average number of seeds per apple. Can you predict the number of seeds by the shape or physical appearance of the apple?

Fig. 1: *Cut an apple in half around the middle.*

Fig. 2: *Scrape out and count the number of viable seeds.*

 TAKE IT FURTHER

- Record the number of robust and shriveled seeds in at least ten apples within one apple variety. Repeat the process with other apple cultivars and compare your results. How does the seed production of organic apples and locally grown varieties compare?

- Flower parts in multiples of five are often a characteristic of *Rosaceae*, the rose family. Closely examine an apple blossom (which is a member of this family). A normal apple blossom consists of five carpels, petals, and styles. How many stamens can you see? Is it a multiple of five?

 PARTS OF AN APPLE

The hypanthium—the tissue that connects the sepals, petals, and stamens—develops into the fruit we eat.

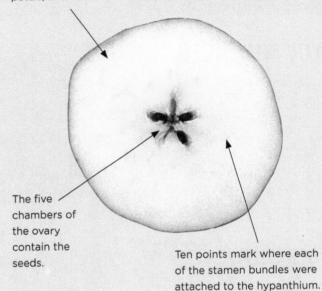

The five chambers of the ovary contain the seeds.

Ten points mark where each of the stamen bundles were attached to the hypanthium.

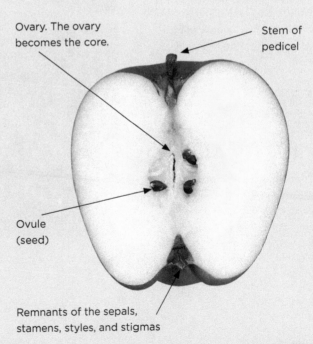

Ovary. The ovary becomes the core.

Stem of pedicel

Ovule (seed)

Remnants of the sepals, stamens, styles, and stigmas

HAND POLLINATE SQUASH

YOU WILL NEED

- blooming squash plant

BEE BUZZ

Although squash is pollinated by honeybees, the squash bee is a squash-pollinating powerhouse that visits only plants in the cucurbit family. Full pollination takes six to eight pollinator visits.

FUN FOR KIDS

Count and record the number of flowers on individual squash plants. Do some plants have more flowers than others?

Members of the *Cucurbitaceae* family (squash, pumpkins, cucumbers, melons, and gourds) have separate male and female flowers on the same plant. Male flowers generate the pollen necessary for the female flowers to produce the fruit. The hand-pollinating technique covered in this lab can be used on all cucurbits.

DIRECTIONS

1. Make sure the plant has both male and female flowers. The male flowers open a week or two before the female flowers in order to begin attracting pollinators. (Fig. 1)

2. Cut off a male flower. Carefully peel away the petals to expose the pollen-covered stamen. (Fig. 2)

3. If the plant was organically grown, instead of throwing the petals away, pop them in your mouth—they're edible!

4. Holding the stamen, gently rub it on the stigma in the center of a female flower. Twirl the stamen around to leave as much pollen as possible on the stigma. (Fig. 3)

5. Visit the hand-pollinated flowers daily. Record your observations with photos.

Fig. 1: *The male squash flowers have long, thin stems that open about ten to fourteen days before the female flowers. There are more male flowers than female flowers on a plant. The female squash flowers have a small fruit (a miniature squash, melon, or cucumber) between the stem and the flower. Female blossoms have shorter stems than the male flowers. If still in doubt, examine the center of the blossoms to identify the stigma, which will be larger than the stamen of the male flower.*

Fig. 2: *Remove the petals of a male squash flower to reveal the stamen.*

 ## TAKE IT FURTHER

- View the squash flowers under a UV light (Lab 26).

- Perform a pollination experiment. Hand pollinate half of the squash plants in your garden and let bees pollinate the rest. Compare the fruit production and quality in each section.

- Harvest male squash blossoms to eat in fresh salads.

Fig. 3: *Brush the stamen of the male flower on the stigma of a female flower.*

POLLINATION JOURNAL

YOU WILL NEED

- notebook
- pen

BEE BUZZ

Honeybees are active at temperatures between 55°F and 90°F (13°C and 32°C).

FUN FOR KIDS

Draw pictures of blooming plants that attract bees. Pay close attention to the number of petals, flower color, and shape. Does the flower have a scent? Do you see other insects on the flowers?

Observe plant blooming cycles and pollinator activity. Record your discoveries in a pollination journal and then compare your results. Fine-tune your senses and let the adventure begin.

DIRECTIONS

1. Take time to walk around your yard, neighborhood, or park a few times every week. Vary the times of day to include mornings, afternoons, and early evenings. Record your observations in a notebook as shown in the chart opposite. (Fig. 1)

2. **Plants and Pollinators.** Choose one plant, or one group of plants, to observe. Count the number of honeybees working the blooms. Return to the same plant at another time of the day. Record your observations on each visit. (Fig. 2)

3. **Bloom Calendar.** Record the date you see the first blooms of a specific plant. Over time, you can compare your observations from one year to the next. For instance, using this chart helped me recognize that dandelions in the area begin to bloom when the temperatures reach about 50°F (10°C). (Fig. 3)

4. **Conclusions.** Examine your observation charts to discover the answers to the following questions:
- What is the earliest time of day you have seen honeybees working?
- Is there a time of day when the honeybees are most active?
- Is there more than one kind of pollinator working a particular plant?
- What types of flowers attract honeybees and other pollinators?
- Do all insects visit flowers at the same time each day?
- What plant has the most or the lengthiest bee activity?

Plant name or description	Date	Time	Weather	Description	Flower color	Flower shape	Number of honeybees	Color of bee pollen (if any)	Other pollinators or insects	Other notes

Fig. 1: *Use the same record-keeping chart for each observation visit to familiarize yourself with the pollinator-friendly plants in your area and to better recognize the patterns of pollinators. Save the charts to aid in making yearly pollination comparisons.*

Fig. 2: *Count and record the number of honeybees on one plant.*

Plant name or description	Date of first blooms	Date of last blooms	Weather description

Fig. 3: *Record the blooming cycle of specific plants in your area*

TAKE IT FURTHER

- Make a virtual pollination journal using a note-taking app such as Evernote that will allow you to take photos of plants, make audio and written notes of your observations, and automatically record the time and location.

- Take a photo of every plant that bees visit in your yard. Identify plants using field guides or consult a nursery.

HEALTH AND BEAUTY

THANKS TO HONEYBEES, THE PRODUCTS OF THE
HIVE OFFER HEALTH AND WELL-BEING FOR ALL.

Every product of the beehive has a purpose and place in the intricate honeybee colony.
Pollen and honey provide bees with protein and carbohydrates for complete nutrition.
Propolis—a resinous substance collected by honeybees from trees—seals the hive,
strengthens comb, and helps keep the hive sterile. Beeswax is transformed into a home,
nursery, and pantry for the bees.

Not only has the humble bee benefited from these products but products of the hive have
also improved the lives of humans for thousands of years. Used externally, honey attracts
and retains moisture, making it perfect for skin-care products such as balms, creams, and
cleansers. Pollen provides a nutritional energy source. The antimicrobial properties of
propolis can contribute to the healing qualities of salves and tinctures. Beeswax, an
emollient, can serve an important role in cosmetics, soothing and softening the skin.

Explore a few of the many beneficial uses of honey, pollen, propolis, and beeswax.

Make. Formulate. Soothe. Heal.

LAB 31

THROAT SOOTHER

YOU WILL NEED

- ½ of a lemon
- mug
- 1 tablespoon (20 g) raw, unfiltered honey
- ginger (optional)
- hot water

Honey has been used for centuries to help soothe the throat and calm a cough. This honey and lemon drink can improve the quality of sleep for both the person under the weather and the rest of the household once the interruption from a cough subsides. If a cough persists, consult your physician.

DIRECTIONS

1. Squeeze the juice from the lemon half into a mug. (Fig. 1)

2. Add the honey and, if desired, a small piece of ginger (Fig. 2).

3. Pour hot water into the mug and stir until the honey is dissolved. (Fig. 3)

4. Slowly sip the warm liquid or cool the mixture and use as a gargle.

BEE BUZZ

A study conducted by Penn State College of Medicine found that honey provides better relief of nighttime coughs than many over-the-counter cold medications.

TAKE IT FURTHER

- Use an herbal-infused honey (Lab 13) to enhance the healing properties of this drink.

- Having trouble sleeping? Before going to bed, place a spoonful of honey under your tongue. Once dissolved, drink 1 cup (240 ml) warm water.

- Make a sage gargle by adding 2 cups (40 g) fresh or (32 g) dried sage to hot water. Steep, strain, and cool. Add 1 tablespoon (20 g) honey, 2 tablespoons (36 g) salt, 2 tablespoons (30 ml) apple cider vinegar, and 1 tablespoon (6 g) cayenne pepper. Gargle with the mixture.

Fig. 1: *Squeeze lemon juice into a mug.*

Fig. 2: *Add honey.*

Fig. 3: *Add hot water.*

FUN FOR KIDS

Make this throat soother for children one year or older and then snuggle under a blanket and read a book together. Here are some suggested titles.

Flight of the Honey Bee, written by Raymond Huber and illustrated by Brian Lovelock

In the Trees, Honey Bees, written by Lori Mortensen and illustrated by Cris Arbo

The Honeybee Man, written by Lela Nargi and illustrated by Kyrsten Brooker

BEEBREAD

YOU WILL NEED

- raw, unfiltered honey
- pollen, fresh frozen or dried
- jar with a tight-fitting lid

BEE BUZZ

Nurse bees ingest beebread to activate the gland responsible for producing royal jelly, a highly nutritious food for larvae and queen development.

SAFETY NOTE

Be aware that some people may have an allergic reaction to bee pollen.

Bees mix pollen and honey to make beebread, a highly nutritious food fed primarily to the brood. It undergoes a type of lactic acid fermentation, making the product more digestible and nutrient rich. This human-made beebread makes the nutrients in pollen more bioavailable to us while maintaining a more nutritionally stable shelf life than pollen alone.

DIRECTIONS

1. The ratio of pollen to honey will vary depending on your preference. Two parts honey to one part pollen is a great place to start. If you prefer more pollen, try using three parts honey to two parts pollen. Fresh frozen pollen is preferred.

2. Pour the pollen into the jar.

3. Add the honey. Tightly secure the lid. (Fig. 1)

4. Place the mixture in a warm, dark location. Turn the jar over three times a day or each time you see pollen at the surface. (Fig. 2)

5. It will take 10 to 14 days for most of the fresh frozen pollen to incorporate with the honey. If using dried pollen, some of the pellets will still be visible. The beebread will be a caramel color with the consistency of frosting.

6. Continue to protect the beebread from light. Either store in a dark place or use amber-colored glass to bottle beebread for yourself or others.

7. Enjoy this delicious treat by the spoonful straight from the jar or add it to smoothies, yogurt, fruit juices, salad dressing, dips, and cookie recipes.

Fig. 1: *Combine pollen and honey in a jar.*

Fig. 2: *Turn the jar over three times a day or when the pollen reaches the surface.*

TAKE IT FURTHER

- If you don't have any beebread, make a quick substitute by soaking a spoonful of pollen in juice overnight. Add it to a smoothie or drink it straight.

- Use beebread when making a honey hydrator (Lab 11).

HEALTH BENEFITS OF POLLEN

- Enhances the immune system with the highest antioxidant activity of any fruit or vegetable, as measured by the ORAC index (oxygen radical absorbance capacity)

- Aids the cardiovascular system with large concentrations of rutin, a compound known to strengthen blood vessels and aid circulation

- Protects against the adverse effects of X-rays

FUN FOR KIDS

Work together to prepare a special treat that children will enjoy making and eating. Combine ¼ cup (65 g) beebread, ¼ cup (65 g) nut butter, ¼ cup (35 g) chopped dried fruit and nuts, and ½ cup (40 g) rolled (old-fashioned) oats. Roll into balls.

PROPOLIS TINCTURE

YOU WILL NEED

- 4 ounces (120 g) propolis, in small pieces
- 2½ cups (600 ml) grain alcohol or vodka, 70 proof (35 percent)
- glass jar with a tight-fitting lid
- coffee filter
- sieve
- small funnel
- dark glass bottles with droppers
- labels

BEE BUZZ

All species of *Apis*, along with many species of stingless bees, collect propolis. The scientific name for the European honeybee is *Apis mellifera*.

SAFETY NOTE

Although rare, some people may have allergies to propolis. Precautions must be taken with alcohol.

One of the best things you can do with the propolis scrapings from the hive is to make a tincture. Alcohol provides the best means to extract the many antibacterial, antifungal, and anti-inflammatory compounds found in propolis. This ratio will make a 20 percent propolis solution.

DIRECTIONS

1. Pour the propolis and grain alcohol into a glass jar with a tight-fitting lid. Do *not* use isopropyl (rubbing, surgical spirit) alcohol! If you do not have a scale, the approximate amount of propolis is a little less than a cup. The exact amounts are not vital if making the tincture for your own use. (Fig. 1)

2. Store the jar in a warm, dark place. A kitchen counter covered with a dish towel is a great place for the tincture to be in view and not forgotten. Shake two or three times daily.

3. After about two weeks, strain the liquid through a coffee filter supported by a sieve. Propolis is very sticky, so avoid using favorite utensils. (Fig. 2)

4. Use a small funnel to fill dark glass bottles with the tincture. (Fig. 3)

5. Immediately label this as 20 percent propolis tincture. Include the date.

6. This tincture will last for years if stored in a cool, dark place.

Fig. 1: *Combine propolis and alcohol in a jar with a sealable lid.*

Fig. 2: *Strain the tincture through a sieve lined with a coffee filter.*

Fig. 3: *Pour the tincture into small glass bottles with dropper lids.*

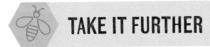

HOW TO USE PROPOLIS TINCTURE

- Boost the immune system by taking a half of a dropper of tincture in a small glass of water or juice during the cold and flu season.

- Dilute a dropper full of tincture with water to use as a mouthwash for mouth sores, dental infections, and gingivitis (gum disease).

- Take the tincture for giardia and intestinal parasites. Of course, see a doctor if dehydrated or if symptoms last more than a week.

- Use topically for minor cuts. Please note that the alcohol does sting.

TAKE IT FURTHER

The highest concentration recommended for making a tincture is a 30 percent propolis solution. If you are interested in a higher propolis concentration, evaporate the alcohol in the tincture to the desired dilution. For instance, evaporate half of the alcohol in this 20 percent solution to make a 40 percent propolis tincture.

PROPOLIS SALVE

YOU WILL NEED

- ½ ounce (14 g) frozen propolis
- coffee grinder
- sieve
- measuring spoons
- 1 ounce (28 g, or 2 tablespoons grated) beeswax (wax cappings preferred)
- 1-pint (500 ml) canning jar
- double boiler
- ¼ cup (60 ml) extra-virgin olive oil
- ¼ cup (60 ml) sweet almond oil
- wooden stir stick or small whisk
- 1 teaspoon honey
- 1 teaspoon propolis tincture (Lab 33, optional)
- 5 drops vitamin E oil
- 5 drops rosemary extract
- 5 drops lavender essential oil
- glass or tin containers
- cheesecloth (optional)

There is a pharmacy in your beehive! Use propolis, an antimicrobial compound that bees collect from plants (mostly trees), to make a healing salve for minor burns and wounds. I have found this salve to be a very effective treatment for cold sores.

DIRECTIONS

1. To make your own propolis powder, grind a few small frozen chunks of propolis in a coffee grinder used only for propolis. If the propolis powder clumps together, sift through a handheld sieve. Measure out 1 tablespoon. (Fig. 1)

2. Put the beeswax in the canning jar and put the jar in a double boiler.

3. Add the olive and almond oils to the beeswax. Remove from the heat. (Fig. 2)

4. Slowly add the fine propolis powder to the wax and oil mixture, stirring vigorously with a wooden stick or small whisk. (Fig. 3)

5. Just as the mixture begins to thicken, add the honey, propolis tincture, vitamin E, rosemary extract, and lavender essential oil.

6. Pour into the glass containers. (Fig. 4)

7. Sometimes the salve may be slightly gritty due the propolis powder. It is fine to use, but if you prefer a very smooth ointment, reheat the mixture and filter through cheesecloth.

BEE BUZZ

After the house bees help the foragers dislodge the propolis from their pollen baskets, the bees mix the propolis with beeswax.

⚠ SAFETY NOTE

Keep wax away from open flames. Unplug heating sources when not in use.

Fig. 1: *Grind frozen propolis into powder.*

Fig. 2: *When the beeswax has melted, add the olive and almond oils to the jar in the double boiler.*

Fig. 3: *Slowly stir in the propolis powder.*

Fig. 4: *Fill containers to continue the cooling process.*

USES OF PROPOLIS SALVE

- minor cuts and wounds
- minor burns
- acne
- athlete's foot
- bedsores
- cold sores

TAKE IT FURTHER

- Make an herb-infused oil using dried calendula flowers, comfrey leaves, and St. John's wort to add even more healing properties to this salve. Place the dried herbs in a clean jar. Make sure both the jar and the lid are completely dry. Cover the herbs with at least 2" to 3" (5 to 7.5 cm) of extra-virgin olive oil. Place in a warm, sunny location for 2 weeks, turning daily to keep the herbs covered. Strain through cheesecloth. Squeeze to extract all the oil.

- Make a quick propolis cream by blending in ½ teaspoon of propolis tincture (Lab 33) for every 1 ounce (28 g) of beeswax cream (Lab 36).

- Explore, in greater depth, the therapeutic benefits of honey when used topically. Honey is antimicrobial, promotes healing, prevents scarring, stimulates new tissue growth, and has been proven to be very effective in treating burn patients.

LIP BALM

YOU WILL NEED

- aluminum foil
- 1 tablespoon (14 g) grated beeswax (wax cappings preferred)
- glass measuring cup
- double boiler
- 2 tablespoons (30 ml) sweet almond oil
- wooden stir stick
- 1 teaspoon lanolin
- 5 drops vitamin E oil
- 8 to 10 lip balm tubes
- paper towels

BEE BUZZ

Almonds, used to make almond oil, are pollinated by honeybees.

⚠ SAFETY NOTE

Keep wax away from open flames. Unplug heating source when not in use.

Make this simple, inexpensive lip balm for your own use or as a gift. You may never buy lip balm from the store again!

DIRECTIONS

1. Prepare a pouring area to catch overflow by folding up and crimping the edges of a piece of aluminum foil. Any lip balm spilled can be saved and reused. (Fig. 1)

2. Melt the beeswax in a glass measuring cup in a double boiler. Glass, stainless steel and ceramic are preferred materials for making skin care products.

3. Add the almond oil and lanolin. Stir with a wooden stick. As soon as the mixture has melted, remove from the heat.

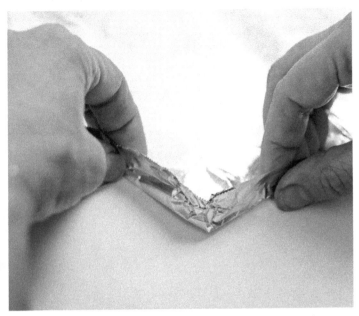

Fig. 1: *Fold up the sides of aluminum foil to catch drips while pouring.*

4. Let the oil and wax cool a bit before adding a few drops of vitamin E oil. Pour into lip balm tubes. It is tempting, but do not hold the tube with your hands. Hot wax will spill onto your fingers! (Fig. 2)

5. Make cleanup easier by wiping the glass measuring cup with a paper towel while still warm.

6. Cap the lip balm once completely cool. Wipe any drips off the tube with a paper towel.

7. Store any lip balm not being used immediately in the refrigerator to extend the shelf life.

Fig. 2: *Carefully pour the mixture into tubes or tins.*

TAKE IT FURTHER

Design and make your own labels on the computer or as individual pieces of art. Protect the label on your lip balm by covering with wide clear tape.

FUN FOR KIDS

Make a variety of lip balms by adding flavors especially made for lip-care products that you can buy online.

BEESWAX CREAM

YOU WILL NEED

Waters

- ³/₄ cup (180 ml) distilled water
- ¹/₂ cup (118 ml) aloe vera gel
- 1 tablespoon (20 g) honey
- 10 drops vitamin E oil
- 1 or 2 drops essential oil of choice

Oils

- ¹/₂ to 1 ounce (14 to 28 g, 1 to 2 tablespoons grated) beeswax
- ³/₄ cup (180 ml) apricot, almond, or grapeseed oil
- ¹/₄ cup (60 ml) coconut oil or cocoa butter

Equipment

- glass measuring cup
- wooden stir stick
- double boiler
- blender
- spatula
- containers

BEE BUZZ

It took the life's work of thirty-six bees to make the honey for this cream.

Combining oil and beeswax (skin protectors) with water and honey (moisturizers) can be a challenge. This recipe will help you do just that. Rejuvenate your skin with this all-natural handmade face and body cream. This is the only cream I use.

Fig. 1: *Combine the distilled water, aloe vera gel, honey, vitamin E oil, and essential oil in a glass measuring cup.*

DIRECTIONS

1. Combine the waters in a glass measuring cup. Stir to blend. (Fig. 1)

2. Melt the oils in a double boiler over low heat. Glass, stainless steel, and ceramic are preferred materials for making skin care products. Once melted, immediately remove from the heat.

3. Pour the oils into a blender. Cool to room temperature. The mixture will be thick and cream colored. (Fig. 2)

4. Use a spatula to scrape the sides of the blender to ensure all the wax is evenly combined in the cream.

5. Put the lid on the blender and turn on to the highest speed. Slowly add the water mixture to the center of the blender while it is running. (Fig. 3)

6. Listen to the blender. Turn it off when it begins to strain. Continue to blend by hand if necessary.

7. Pour the mixture into containers. Store the cream not being used immediately in the refrigerator to extend the shelf life. (Fig. 4)

Fig. 2: *Once the oils are melted, pour into a blender and let cool to room temperature.*

Fig. 3: *Turn the blender on and slowly add the waters.*

Fig. 4: *Pour into containers.*

TAKE IT FURTHER

Try using an herb-infused oil, such as calendula (Lab 34).

FUN FOR KIDS

This recipe blends oil and water. What happens when you drop oil into water or water into oil?

HONEY AND POLLEN CLEANSER

YOU WILL NEED

- ½ cup (120 ml) water
- kettle
- 1 green tea bag
- glass jar with a tight-fitting lid
- 2 teaspoons bee pollen, fresh frozen or dried
- 2 tablespoons (30 ml) aloe vera gel
- 2 tablespoons (40 g) honey
- blender (optional)
- small bottles (optional)

BEE BUZZ

Pollen pellets carried on the back legs of a honeybee may contain several thousand pollen grains. Some may have close to a million, depending on the size of each grain.

As a facial wash, this cleanser is both nourishing and gentle to the skin. The antibacterial properties of honey and the anti-inflammatory effects of pollen are both excellent for acne-prone skin.

DIRECTIONS

1. Boil the water in a kettle, then let it cool for a few minutes. Place the green tea in a glass jar, pour the hot water over, cover, and let steep for 3 to 5 minutes. Remove the tea bag and let cool for about 5 minutes. (Fig. 1)

2. Add the bee pollen (fresh frozen is preferred), aloe vera gel, and honey to the tea. (Fig. 2)

3. Tightly screw on the lid. Shake the jar to blend all the ingredients. Mix in a blender if using dried pollen. (Fig. 3)

Fig. 1: *Make ½ cup of green tea in a glass jar.*

Fig. 2: *Add the bee pollen, aloe vera gel, and honey to the jar.*

4. Transfer to smaller bottles. Use this mixture in the next few days. Store the cleanser not being used immediately in the refrigerator to extend the shelf life to one week.

5. Use this gentle cleanser daily. Shake and then massage on your face in circular motions using your fingertips. Remove with a warm cloth. Rinse with cool water.

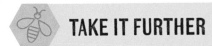

TAKE IT FURTHER

In a pinch, make a quick cleanser using 1 tablespoon (15 ml) milk, 1 tablespoon (20 g) honey, and 1 teaspoon ground oatmeal.

Fig. 3: *Seal and shake vigorously.*

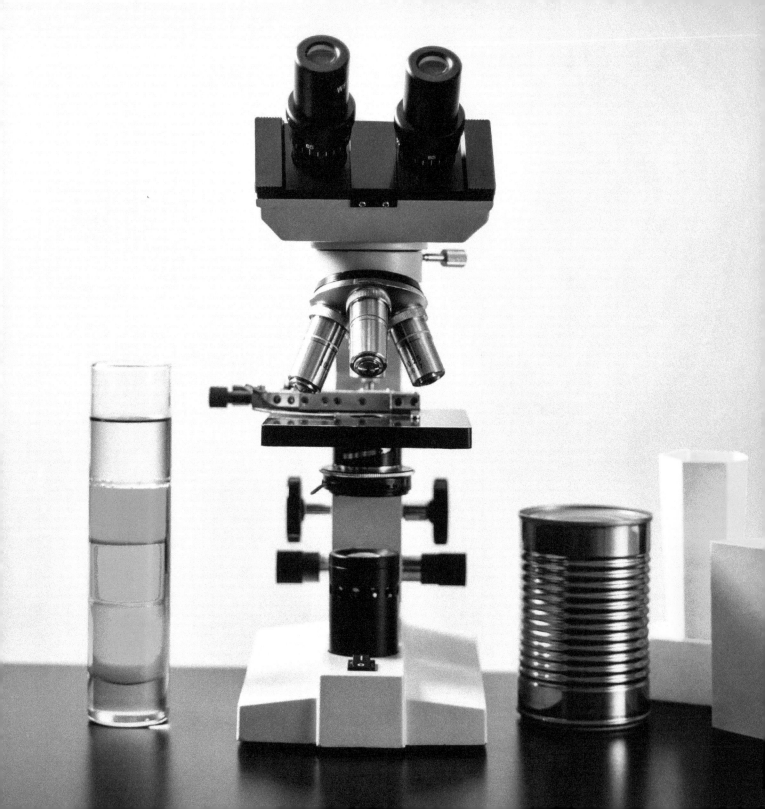

EXPERIMENTS

HONEYBEES OPEN THE DOOR TO WONDER, WHICH ENCOURAGES CURIOSITY.

The world of honeybees is complex and utterly fascinating. Tell someone you keep bees and the questions begin in rapid succession. Why is the bee population declining? What can I do the help the bees? How do bees make honey? My first question as a beekeeper was, "How do thousands of individual bees work together as one?" Learning and discovery often begin with curiosity.

The social structure of the hive captured my imagination and filled me with awe many years ago. It still does. Each bee larvae receives approximately 10,000 visits from nurse bees during development. How is that possible? With each new fact I learn about bees, I find I want to learn more.

Exploration is part of that learning process. How does honey compare with other sweeteners? What do bee legs look like under a microscope? Can bees recognize shapes and colors? How do bees make a buzzing sound?

Examine. Discover. Make. Listen.

HONEY DENSITY

YOU WILL NEED

- clear glass container (tall and thin works best)
- sugar
- saucepan
- spoons
- 2 jars or cups
- food coloring
- light-colored corn syrup
- honey
- vegetable oil

The size of a liquid's molecules and the amount of space between them will determine the density of a liquid. Honey has big molecules jammed together, while the molecules in vegetable oil are spaced far apart. Compare the density of three sweeteners. The denser the liquids, the heavier they will be.

DIRECTIONS

1. The size of the glass container will dictate how much of each liquid will be needed.

2. Make a small amount of thick sugar syrup by dissolving two parts sugar to one part water in a saucepan on the stove. Stir constantly. This is the sugar syrup ratio fed to bees with low food stores in the fall.

3. Transfer to a jar and then stir in a few drops of red food coloring. Place in the refrigerator for 5 minutes to cool down.

4. Pour some of the corn syrup into another jar. Add a few drops of green food coloring and stir. (Fig. 1)

5. Now for the fun part. Fill a glass container less than a fourth full with honey. (Fig. 2)

6. Slowly add the colored corn syrup. (Fig. 3)

7. Next, pour in the sugar syrup. (Fig. 4)

8. Last of all, top it off with vegetable oil to make another density comparison.

9. Which sweetener is the densest?

BEE BUZZ

Honey is about 36 percent denser than water.

Fig. 1: *In a jar, blend green food coloring into the corn syrup.*

Fig. 2: *Pour a layer of honey into the glass container.*

Fig. 3: *Pour a layer of colored corn syrup on top of the honey.*

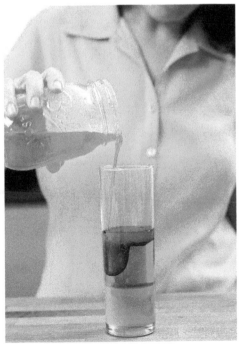

Fig. 4: *Slowly add the colored sugar syrup before adding the vegetable oil.*

TAKE IT FURTHER

- Using a scale, separately weigh the same volume of each liquid. Record your findings. Repeat the experiment. How does the weight of each liquid compare to the density experiment?

- Try adding other liquids, such as milk, to make more layers.

- What happens when you drop a small item like a paper clip, penny, almond, or grape into the jar?

FUN FOR KIDS

Add the honey, corn syrup, and oil to a glass container. Shake or stir it up. What happens? Let it sit for at least 20 minutes and look at it again.

COMB STRENGTH

YOU WILL NEED

- ruler
- pencil
- 4" × 6" (10 × 15 cm) index cards
- tape
- heavy objects (cans of food work well)

The hexagonal structure of bee comb is both strong and an efficient use of space. Engineers have used the architectural structure of the hexagon to strengthen airplane wings and satellite walls. Fold card stock into "cells" with three different shapes—square, triangle, and hexagon—and use weights to test the strength of each shape.

DIRECTIONS

1. Use a ruler and pencil to mark the folding points for each shape on the top and bottom edges of an individual index card. Space the marks as indicated below.
   - Triangle: 2 marks 2" (5 cm) apart to make 2 folds and 3 equal sides
   - Square: 3 marks 1½" (3.8 cm) apart to make 3 folds and 4 equal sides
   - Hexagon: 5 marks 1" (2.5 cm) apart to make 5 folds and 6 equal sides

2. Draw a line from the top mark to the corresponding bottom mark to use as a folding guide. (Fig. 1)

3. Carefully fold along each line.

4. Tape the open edges together to form the different shaped tubes. (Fig. 2)

5. Stand the tubes on their ends. Test their strength by placing the same amount of weight on the top of each. (Fig. 3)

6. After the first round, continue adding weight to see which tube is the last one standing.

BEE BUZZ

One pound (454 g) of beeswax comb will hold about 22 pounds (10 kg) of honey. Honeybees also strengthen the beeswax for comb building by adding propolis.

Fig. 1: *Mark folding lines to create equal sides for each shape: triangle, square, and hexagon.*

Fig. 2: *Construct the tubes by connecting the short edges together using tape.*

Fig. 3: *Begin the strength test by placing the same amount of weight on each tube.*

TAKE IT FURTHER

- Can you build a paper honeycomb structure that will hold your own weight?

- What happens when you vary the length and diameter of each shaped tube?

FUN FOR KIDS

Instead of folding paper, try cutting up paper towel tubes and taping them together to create the cells in a hive. Stuff each "cell" with colored tissue paper to represent the larvae (white), pupa (brown), and pollen (yellow).

LAB 40

HONEYBEES UP CLOSE

YOU WILL NEED

- worker honeybee
- jar with lid
- dissecting scope or compound microscope
- slides and slide covers
- tweezers or forceps
- glycerin (optional)
- notebook
- pen

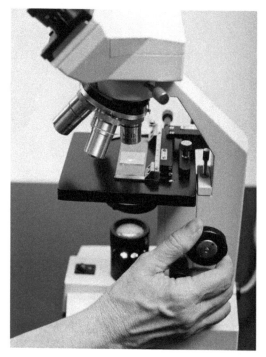

Use magnification to gain a greater understanding and appreciation of the anatomy of a worker honeybee.

DIRECTIONS

1. Go out to the hive and look for a lifeless bee near the entrance to examine. If no bee is found, use a jar to capture a honeybee at the hive entrance or on a flower. Put the bee in the freezer for at least an hour.

2. **Hair.** Begin your bee anatomy examination by placing the bee on a dissecting scope or slide. Look at the hair on the body of the bee. What purpose does the hair serve?

3. **Head.** Locate the compound eyes. You may be able to see the three simple eyes called ocelli on the top of the head. Where on the head are the antennae located? Can you see the proboscis (strawlike tongue) and mandibles (mouth parts)?

4. **Hind leg.** Using two tweezers, detach the hind leg as close to the thorax as possible. The leg can be viewed directly on a slide or add a drop of glycerin first before placing the leg on the slide for greater clarity. The largest of the nine segments that can be easily seen are the femur, tibia, and basitarsus. The foot consists of two claws and an adhesive pad. (Fig. 1)

5. **Pollen-collecting devices.** Locate the pollen basket on the outer surface of the tibia. It is fringed with long, curved hairs. The pollen press is situated in the joint of the tibia and the basitarsus. You may be able to see the pollen rake used to comb pollen off the basitarsus into the pollen press. When the worker bee straightens her leg, the pollen is forced up into the pollen basket. (Fig. 1)

6. **Foreleg.** Remove the foreleg and place it on a slide. Locate the notch of the antenna cleaner on the basitarsus. The small clasp on the tibia may also be visible. (Fig. 2)

7. **Wings.** Detach one set of wings using tweezers or forceps. Separate the larger forewing from the smaller hindwing and place on a slide. View the veins that strengthen each wing and divide the

BEE BUZZ

Honeybee antennae are made up of thousands of sensory organs to help the bee smell, touch, taste, and detect sound. A worker bee has 3,600 to 6,000 sensory organs on each antenna while the drones have 30,000.

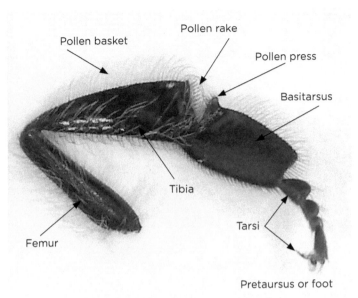

Fig. 1: *Hind leg of a worker bee*

Pollen basket

Pollen rake

Pollen press

Basitarsus

Tibia

Tarsi

Femur

Pretaursus or foot

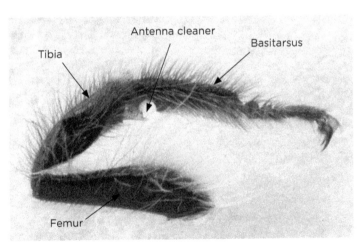

Antenna cleaner

Basitarsus

Tibia

Femur

Fig. 2: *Foreleg of a worker bee*

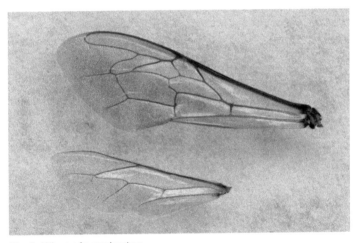

Fig. 3: *Wings of a worker bee*

wing into cells. These veins and cells are important for identifying bee species. Are the wings frayed or worn from flying? (Fig. 3)

8. **Abdomen.** You may be able to see six of the nine segments. Locate the last four visible segments on the underside of the abdomen where the wax is secreted. Gently squeeze the abdomen from the sides to view the stinger.

9. Make drawings and notes in your notebook. Thank the bee for giving its life for entomological research.

CAN BEES READ?

YOU WILL NEED

- sugar
- water
- saucepan
- blue marker
- 5 index cards, 3″ × 5″ (7.5 × 12.5 cm)
- 5 sealable plastic bags
- 5 identical shallow plates
- stones for weights
- corks (optional)
- small notebook

BEE BUZZ

Honeybees, like most insects, not only have a brain in their head but also several sub-brains or ganglia spread throughout their bodies. There are two ganglia in the thorax and five ganglia in the abdomen that send and receive messages.

Put a bee's memory and ability to recognize shapes to the test with this simple experiment best done in the early spring.

DIRECTIONS

1. Make nectar for the bees by dissolving one part sugar in one part water using a saucepan on the stove. Let the mixture cool to room temperature.

2. In the meantime, with a marker, draw a different shape, big and bold, on each index card. Avoid using shapes that are too similar, such as a solid circle and solid square. (Fig. 1)

3. To protect the cards from the weather, place each one in a separate sealable plastic bag.

4. Find a flat, covered spot outside where the experiment will not be disturbed. A table with an umbrella is perfect.

5. Keep the cards from blowing away by placing small stones, sand, or dirt in the bag under the index card. Lay on a table about 2′ (60 cm) apart. (Fig. 2)

6. Set a plate near each card. Plates should have a rim for the bees to stand and drink from. Pour the sugar syrup onto one plate, making note of the corresponding shape, and plain water in the other four. If needed, float corks in each plate for the bees to drink from without drowning. Change the sugar water every seven days. (Fig. 3)

7. Write down daily observations. How many days until bees find the sugar syrup? After you've seen bees at the sugar syrup plate for a few days, switch that plate with a plain water plate. How does that change affect the bees?

FUN FOR KIDS

Paint or draw a symbol on each of your beehives to help the bees find their home.

Fig. 1: *Using a marker, draw a different symbol on five index cards.*

Fig. 2: *Anchor the index cards by placing small stones in the bag under the drawn shape.*

Fig. 3: *Pour sugar water onto one plate.*

TAKE IT FURTHER

- Take away the symbols, leaving only the feeding plates. Mix different concentrations of sugar syrup to pour onto each plate. Which solution do the bees prefer?

- Are any of the feeding bees from your hive? Lightly sprinkle powdered sugar on the feeding bees. Quickly go to your hive, sit near the entrance, and see if any of the bees covered in sugar are returning to your hive.

- Lay on your back near the feeding station experiment. Try to watch which direction the bees are flying. Spotting the bees is easier on a cloudy day. They will fly in a circle before taking off. Are they flying in the direction of your hives?

BUZZ OF BEES

YOU WILL NEED

- earbuds and microphone for a smart phone
- recording device, such as a smart phone, iPad, tablet, or laptop computer
- scissors
- a hive of bees

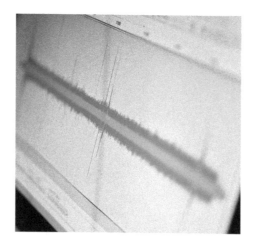

BEE BUZZ

The movement of bees' wings creates sound, but sound is also produced when bees uncouple their wings and flight muscles to generate heat.

FUN FOR KIDS

- A honeybee can move its wings about 11,400 beats per minute. Count how many times you can flap your "wings" in one minute. How does that compare to the wing beats of a honeybee?

- Play a buzzing communication game. Hide a paper bee somewhere in the house or yard for friends to find. Communicate the bee's location by making louder buzzing sounds the closer someone gets to the hiding place.

The sounds that bees make as part of their complex communication system can tell us many things, such as when a colony is upset or queenless. Not all sounds produced by bees are within our hearing range. But as we know from experience, the sounds we can hear in a hive vary. Record the buzz of bees by using readily available equipment.

DIRECTIONS

1. Use a headset from a smart phone to make an inexpensive external microphone for your recording devise. With scissors, cut off the earbuds, leaving the microphone segment for recording. (Fig. 1)

2. Use the voice memo app on your phone or download other recording apps such as CinixSoft or Hokusai. Test your equipment and recording software before heading to the hive.

3. Place the recording device near the hive to capture the buzz of bees. Hang the microphone in front of the hive entrance. Press record. Either let the recorder run while doing your hive inspection or stop the recording after a few minutes. (Fig. 2)

4. Immediately rename the file "hive entrance" or any other reference note to help identify the recording.

5. Make another recording to capture sounds inside the hive. Open the hive and lower the microphone between the frames into the brood chamber. (Fig. 3)

6. Record the hive at different times of the day and under different circumstances to discover any noticeable sound variations.

Fig. 1: *Create an external microphone for a recording device by cutting off the unneeded earbuds.*

Fig. 2: *Place the recording device at the entrance of the hive.*

Fig. 3: *Lower the microphone into the hive.*

Fig. 4: *Analyze the frequency of the buzzing of bees using the free software Audacity.*

 ## TAKE IT FURTHER

- Sound waves are measured by their frequency in Hertz (Hz), or cycles per second. The most common frequencies in the hive range from 190 Hz to 250 Hz. Audacity, a free audio editor, can analyze the frequencies of the captured sounds. Import the audio files to your computer and then open them in Audacity. Analyze the files by using plot spectrum and then log frequency to display a graph. What is the dominant frequency in your recordings? (Fig. 4)

- Compose a piece of music using the sound of the bees and other recorded music tracks. What musical notes correspond to the frequency of the buzz of bees?

- Use one of the recordings as a ring tone for your phone.

- Capture video and audio by recording the bees at the hive entrance.

UNIT 7

ART

THE FORM AND FUNCTION OF A HONEYBEE HIVE INSPIRES CREATIVE EXPRESSION.

Honeybees have been the inspiration for poetry, songs, and artwork for millennia. Henry David Thoreau, Kahlil Gibran, Emily Dickinson, and other writers and poets have found their muse in bees. Shakespeare wrote in *Henry V*, "For so work the honey-bees, creatures that by a rule in nature teach the act of order to a peopled kingdom." Artists' representations of bees are found in paintings, sculptures, and art from around the world: stamps from Russia, ancient coins from Greece, and the coat of arms of Napoleon I from France. Honeybees, and all they represent, speak a common language through art worldwide.

May bees inspire your creativity. Be fearless. Find your muse. Capture the beauty and symmetry of hexagonal cells through comb rubbings. Follow in the footsteps of ancient Egyptians by adding pigment to melted beeswax. Collaborate with bees by placing artwork in the hive. Brighten up beehives just begging for color, lines, and designs.

Create. Collaborate. Decorate.

LAB 43

COMB RUBBINGS

YOU WILL NEED

- cotton fabric (apron, tote bag, or napkins)
- plastic foundation
- clipboard
- beeswax art blocks (Lab 44) or crayons
- iron and ironing board
- paper towels

FUN FOR KIDS

Create beeswax comb rubbings using paper instead of fabric.

The hexagonal shape of beeswax comb lends itself to artistic creation. Unlike actual beeswax comb, plastic foundation is durable, easy to clean, and a consistent height, making it perfect for fabric rubbings. An apron is used for this project.

DIRECTIONS

1. Practice the process using a piece of scrap fabric before beginning the actual project.

2. Prewash the fabric to be used.

3. Place the section of fabric to be colored on top of the plastic foundation. (Fig. 1)

4. Slide the fabric and the plastic foundation under the clip of the clipboard to help anchor the fabric. (Fig. 2)

5. Use a beeswax art block or crayon to apply color to the fabric using firm, one-way strokes while holding the fabric in place. Avoid back-and-forth strokes, as this can sometimes shift the fabric. (Fig. 3)

6. Unclip and arrange an additional section of fabric to color on the plastic foundation. Another option is to slightly move the colored fabric and add a different colored layer on top of the first layer. The distinct hexagonal shapes will no longer be visible, but the effect can add interest.

7. After you have finished coloring all sections of fabric, sandwich the colored fabric between a few layers of paper towels. Iron the colored fabric to remove the wax while retaining the pigment. (Fig. 4)

8. Depending on the fabric and the crayons used, finer woven fabric often retains the color better. Some colors on coarser fabrics may wash out. If possible, test a small piece of the fabric before hand washing in cold water.

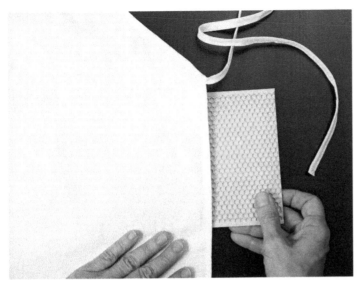

Fig. 1: *Place a section of fabric on top of the plastic foundation.*

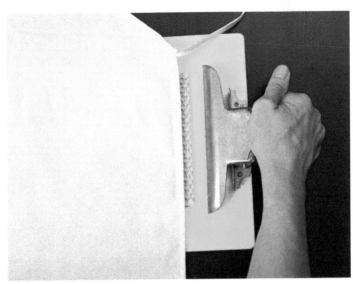

Fig. 2: *Hold the fabric and plastic foundation in place with the clip of a clipboard.*

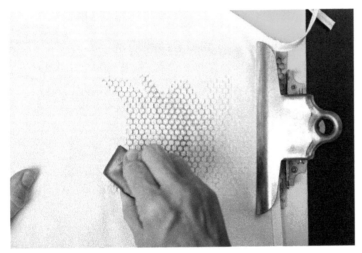

Fig. 3: *Apply color in one direction using a beeswax art block or crayon.*

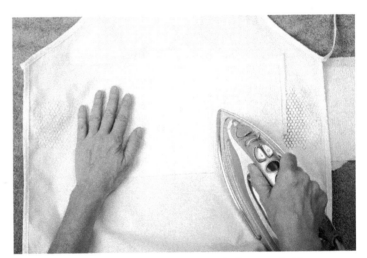

Fig. 4: *Iron between paper towels to remove the beeswax but leave the pigment.*

🐝 TAKE IT FURTHER

- Make a stencil by cutting out a shape in the middle of a piece of card stock. Make the rubbing inside the stencil.

- After applying the crayon color, use acrylic paint to add contrasting images and details to your creation.

BEESWAX ART BLOCKS

YOU WILL NEED

- newsprint or cardboard
- double boiler
- carnauba wax (1 part)
- white beeswax (3 parts)
- measuring spoons
- powdered pigments
- small (3 ounce, or 84 g) paper cups
- wooden craft sticks
- ice cube tray (optional)
- ruler
- marker

BEE BUZZ

The optimum temperature inside the hive for bees to secrete wax is 91°F to 97°F (33°C to 36°C).

⚠ SAFETY NOTE

Keep wax away from open flames. Unplug heating source when not in use.

Make your own art blocks using a mixture of beeswax, carnauba wax, and powdered pigment. You will need about 1 ounce (28 g) total wax for each art block.

DIRECTIONS

1. Set up a workstation with newsprint or cardboard to protect your work surface. Set up a double boiler (a clean can in a saucepan works well). Melt the carnauba wax and beeswax in the can in a ratio of one part carnauba wax to three parts beeswax.

2. Measure 2 teaspoons of powdered pigment into a small paper cup. (Fig. 1)

3. Pour the melted wax mixture into the cup until it is half to three-quarters full. The ratio is 2 teaspoons of pigment to 2 tablespoons (1 ounce, or 24 g) wax. The more pigment, the more concentrated the color. Earth pigments work very well. (Fig. 2)

4. Place the wax back on the heat source and reserve.

5. Immediately stir the mixture using a wooden craft stick until blended.

6. For square art blocks, pour the mixture into an ice cube tray. Square corners allow for greater versatility. Or you may want to keep it simple by allowing the wax to harden in the paper cups. (Fig. 3)

7. Repeat steps 2 through 6 for each different colored art block.

8. Once completely cooled, snap the art blocks out of the ice tray or peel the paper cup off the hardened art block. If you are having difficulty removing the art blocks, place the ice cube tray in the freezer for a few hours and try again.

9. Let the artwork begin!

Fig. 1: *Measure 2 teaspoons of powdered pigment into a paper cup.*

Fig. 2: *Cover the powdered pigment with the melted wax mixture.*

Fig. 3: *After stirring the mixture, pour into an ice cube tray, if desired.*

Sample packets of cosmetic colors offer an inexpensive way to make a variety of colors.

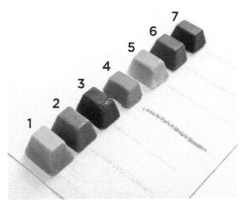

Experiment with different pigments.
1) turmeric, 2) paprika, 3) beet powder,
4) water oil colors, 5) pastels, 6) earth pigment,
7) cosmetic pigment

FUN FOR KIDS

Use food-safe pigments like turmeric and paprika to make these art blocks.

TAKE IT FURTHER

- These beeswax art blocks are perfect for making comb rubbings (Lab 43) or encaustic cards (Lab 45).

- Make quilt squares with beeswax art blocks. Draw a picture on cotton fabric, pressing as hard as possible with the beeswax art blocks. A piece of freezer paper (wax side up) or sandpaper under the fabric will prevent slipping and provide texture to the artwork. Remove the wax but leave the pigment by ironing between layers of paper towels.

- Experiment with changing the ratios of beeswax and carnauba wax. Try adding talc or white powdered clay to the mixture.

ENCAUSTIC CARDS

YOU WILL NEED

- newsprint or cardboard
- electric griddle with low heat setting or warming tray
- aluminum foil
- tape
- garden gloves
- colored beeswax blocks
- white 4" × 6" (10 × 15 cm) index cards
- paper towels
- scissors or paper cutter
- glue stick
- colored paper or card stock

BEE BUZZ

Bees twelve to eighteen days old have the best developed and most productive wax glands.

⚠ SAFETY NOTE

If using a heat source that is hotter than a warm setting, take great care to protect your hands.

If you tend to be intimidated by the visual arts, this is the project for you. Everyone can create beautiful artwork with this simple encaustic monotype technique.

DIRECTIONS

1. Set up a workstation with newsprint or cardboard to protect your work surface. Cover an electric griddle or warming tray with aluminum foil. Secure with tape. Turn to the very lowest heat setting available.

2. Put on a pair of garden gloves to protect your hands from the heat. Hold the beeswax blocks on the griddle surface to melt small pools of colored wax onto the foil. (Fig. 1)

3. Place an index card on the melted wax. Rub the paper with your gloved fingers or a folded-up paper towel to allow the index card to have full contact with the wax. (Fig. 2)

4. Either lift or peel the paper away from the griddle. It is difficult to plan the end result, but the random beauty is sure to delight. (Fig. 3)

Fig. 1: *Melt beeswax blocks on aluminum foil placed on the surface of a griddle turned to the lowest setting.*

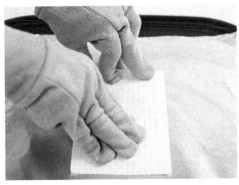

Fig. 2: *Press an index card onto the melted wax.*

Fig. 3: *Peel the index card off the griddle.*

5. Repeat with another piece of paper for a lighter print. Clean the foil with a paper towel before beginning the next print. Replace the foil when it becomes "muddy."

6. Trim the edges of the paper to the desired shape and size. Glue to a folded piece of card stock to make a unique and colorful card. (Fig. 4)

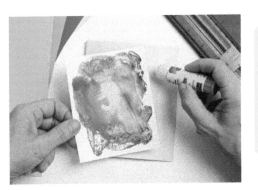

Fig. 4: *Create a card by gluing onto folded card stock or paper.*

FUN FOR KIDS

Use colored beeswax blocks directly on paper without heat.

 ## TAKE IT FURTHER

- Make your own colored beeswax blocks (Lab 44) or purchase encaustic medium blocks at an art supply store.

- Place the card stock or paper on the griddle first. Melt the wax directly on the paper.

- Instead of paper, press or roll other items onto the melted wax. Try using white fabric, canvas, cardboard, corrugated plastic, or glass.

- Make simple encaustic medium directly on the griddle using repurposed mint tins to mix pigment and white wax. Shorten the handles of small paintbrushes to better fit in the tins. (Fig. 5)

- If you would like to try a more traditional encaustic technique of layered beeswax on wood, melt eight parts beeswax and then add one part damar resin to increase hardness and luster. Do not heat above 220°F (104°C).

Fig. 5: *Use mint tins to mix pigment and wax.*

BEES AS ARTISTS

YOU WILL NEED

- a hive of bees
- base material such as plastic foundation, plywood, corrugated plastic, Plexiglas, Lucite, or Masonite
- tools to cut the base material
- beehive frames
- card stock
- art supplies
- glue
- beeswax
- double boiler
- brush or sponge

BEE BUZZ

Hundreds of honeybees contribute to the construction of a single cell of the comb.

FUN FOR KIDS

Place an old toy or another object in the hive for the bees to transform into art.

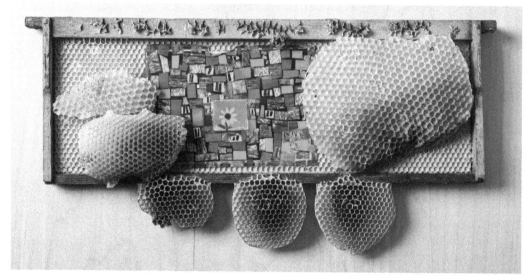

A colored copy of a paper mosaic created by Doris Bickley was glued onto a beekeeping frame. The bees built comb on the bottom of this medium frame placed in a deep super. Previously built comb was then attached to the frame using hot wax.

Collaborate with the bees to create unusual works of art. You provide the artwork, bees add the comb. This project needs to be done in the early springtime or during a honey flow when bees are building comb.

DIRECTIONS

1. First you will need a base for your artwork that will withstand the humidity in the hive. You can use plastic foundation or cut the base material to fit into a deep, medium, or shallow hive frame.

2. Create your artwork on card stock using markers, paints, pens, crayons, charcoal, or collage materials. You can place your original work in the hive or use colored copies of the artwork.

3. Glue your piece onto the base material. Leave the other side blank to add artwork once the bees have done their part. (Fig. 1)

4. Melt the wax in a double boiler. Brush the entire surface of both sides with a very thin coat of melted wax (Lab 3). (Fig. 2)

Fig. 1: *Glue artwork onto the base.*

Fig. 2: *Cover both sides with a thin layer of beeswax.*

5. Scrape off some of the wax if it is obscuring the artwork.

6. Place the frame in the middle of a super.

7. Check every few days to see what the bees are up to. Sometimes they will build wax quickly, other times it may take weeks before the bees venture into comb building on the frame.

8. Experiment with different techniques and ideas to create with the bees.

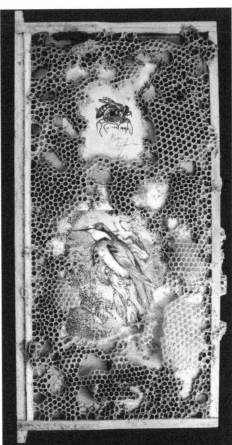

Ladislav Hanka created Bee Eater, *an etching with aquatint and drypoint. The wood engraving of a bee was glued down into a beekeeping frame along with an etching of a bee eater and inserted into a living beehive. There the bees chewed out parts of the paper and added their own contributions of wax done in patterns of their own choosing.*

While relocating a swarm of bees found in an owl nesting box, Anna Gieselman discovered this unique piece of art entirely crafted by honeybees.

 TAKE IT FURTHER

• Take a closer look at your beekeeping equipment. Has beeswax comb transformed your inner cover into a piece of art?

• Create a collage using photographs, artwork, burr comb, pollen, and beeswax.

PAINTED HIVES

YOU WILL NEED

- pencil
- paper
- exterior latex or acrylic paints
- paintbrushes in a variety of sizes
- polyurethane (optional)

BEE BUZZ

Beekeepers in Slovenia have a long history of beautifully painted hives going back over 250 years.

FUN FOR KIDS

Instead of exterior paint, use nontoxic markers, tempera paint, chalk, crayons, or even mud for finger painting. Cover with a clear water-based sealer or repaint when weathered.

Beekeeper and bee-centric jewelry artist Anna Gieselman created this hive to celebrate the flower-bee connection.

Beehives are a blank canvas just waiting for inspiration to strike. Pick up a paintbrush and jazz up your apiary with one-of-a-kind creations.

DIRECTIONS

1. **Design.** Brainstorm design concepts for your hive. Things to consider:

- Will your hive have a theme?

- What images or subjects would you like to work with?

- Do you want to make a statement or express a thought or an opinion?

- Would you like realistic artwork or an abstract design?

- Do you prefer to paint random and spontaneous shapes, colors, and patterns?

- Will it be a group or family project, with each person designing and painting one panel of the hive?

2. Planning.

- Be bold, big, and colorful with your design. Details are difficult to see from a distance.

- Sketch many different designs for the art hive on paper. Do not spend time on details. Pick out your favorites to further develop.

- Choose the colors you will need. Black is great for outlines. Primaries (red, blue, and yellow) make it possible to mix your own colors.

3. Do it!

- Completely cover the hive with a base layer of exterior latex paint.

Fig. 1: *Use a pencil to sketch your design onto the bee super.*

Fig. 2: *Apply paint to the hive.*

- Lightly sketch the design on the hive using a pencil. Do not worry about putting the lines on perfectly. (Fig. 1)

- Draw with your full arm, not just with your fingers.

- If you are painting the outlines in black, do that first. Allow the paint to dry. Go back and fill in with color. Acrylic paints dry a little slower and contain more pigment for better coverage. Exterior latex is much cheaper.

- Paint away! If using acrylics, apply a protective polyurethane coat to the finished painting. (Fig. 2)

This hive, painted by Marjorie Molnar, is inspired by Bauernmalerei, a decorative painting style often referred to as Bavarian folk art.

 TAKE IT FURTHER

- Make a collage by gluing photos and pictures onto the hive. Cover with a protective polyurethane finish.

- Promote artistic expression and honeybee awareness at a public event. Supply permanent markers for people to decorate a hive or write a message of gratitude to the bees or a beekeeping advocate.

- Have a "Wild and Wonderful Beehive Tour" by displaying hives painted by local artists in public places, such as the library, city hall, grocery store, farmers' market, or other local businesses. Auction off the hives or cast monetary votes as a fundraiser.

UNIT
8

SAVE THE BEES

THE ALTRUISM IN HONEYBEES BRINGS OUT THE BEST IN US.

A honeybee's actions are not performed for her individual benefit, but for the well-being of the entire hive. This aspect of honeybee behavior inspires me to step out of myself and look at the world from a bigger perspective. What can I do, not for myself, but to help make the world a better place?

Let's face it—pollinators need all the help and support they can get. Between pesticides, pathogens, poor nutrition, decreased forage, and habitat loss, the odds are stacked against them. Nature is resilient, but let's give a little boost to our pollinators when and where we can.

Provide shelter for native pollinators, plant a pollinator smorgasbord, spread pollinator awareness among the public, and take action in small and large ways to benefit not only the bees but also the world.

All we are saying is give bees a chance.

Build. Plant. Talk. Celebrate. Advocate.

MASON BEE HOUSE

YOU WILL NEED

- parchment paper
- scissors
- pencil
- tape
- ruler
- metal can
- wire cutters
- wire

BEE BUZZ

Worldwide there are about 200 species of mason bees, with around 140 species found in North America.

FUN FOR KIDS

In the fall, conduct an experiment by investigating one of the tubes. Carefully unwind the paper tube to reveal the cocoons, mud (gray), pollen (yellow), mites (red/orange), and bee poop (brown/black).

Mason bees are solitary bees that pollinate many crops, including almond, apple, cherry, pear, and plum. Many are slightly smaller than the honeybee and are typically metallic blue or blue-black in color. Create a comfortable condo out of paper to protect and shelter these unassuming pollinators.

DIRECTIONS

1. Tear a strip of parchment paper about 5" (12.5 cm) wide. With scissors, cut the strip into squares. Parchment paper decreases the chances of mold growth. (Fig. 1)

2. Place a pencil in the corner and roll diagonally into a tube. (Fig. 2) Secure with one piece of tape around the center. (Fig. 3) Remove the pencil.

3. Close one end by folding over the point and taping.

4. Cut off the other end to make a 5" to 6" (12.5 to 15 cm) long tube. Vary the length of the tubes. This will help the bees better identify their home.

5. Continue making tubes. Depending on the circumference of the can, you will need between 50 and 70.

6. With wire cutters, cut two pieces of wire long enough to encircle the can. Twist a loop in the middle of each piece of wire before tightly wrapping the wire around the can. (Fig. 4)

7. Fill the can with the open end of the tubes facing out.

8. Hang outside and help the bees. In the spring, hang a new house next to the old one for emerging bees.

HANGING TIPS

To ensure your mason bee house will not sway, attach it firmly under the eaves of your house or to the side of a shed, fence, or tree. Bees prefer a sheltered spot out of the wind facing east or south at least 3' (1 m) off the ground. Hang the bee house as close to flowering plants as possible. Do not spray insecticides on or around your bee house.

Fig. 1: *Cut parchment paper into squares.*

Fig. 2: *Roll paper around a pencil diagonally.*

Fig. 3: *Tape in the center to secure.*

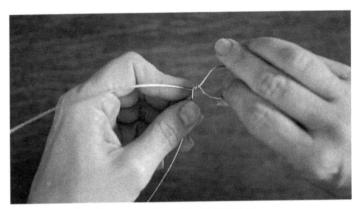
Fig. 4: *Twist a hanging loop in the middle of each wire.*

TAKE IT FURTHER

- Leave areas of bare ground in the yard or dig up dirt for bees to make into mud for sealing the holes.

- In the fall, harvest the hibernating cocoons from the paper tubes to take your mason bee rearing to another level. Destroy wasp- and mite-infested cocoons (lighter colored, crunchy ones), wash the healthy ones (dark gray and firm) and then store the cocoons in a cool place until the spring.

- Make extra mason bee houses to hang throughout your community. Approach a local park, business, religious institution, or city department about making and installing solitary bee houses on their property.

ABOUT MASON BEES

The mason bee places a small ball of pollen and nectar in the back of the tube and then lays an egg. Next, she makes a mud divider, repeats the pollen and nectar process, and lays another egg. She does this over and over until she fills the tube, and then she seals the entrance with mud. One tube holds five to ten eggs. Mason bees will emerge in early spring when temperatures consistently reach 55°F to 60°F (13°C to 15.5°C).

SEED BALLS

YOU WILL NEED

- mixing container (dish basin, foil pan, or cardboard box lid)
- 5 parts compost
- 1 part seeds (nectar- and pollen-producing plants appropriate for your area)
- 3 parts powdered clay
- water
- sprayer
- newspaper

BEE BUZZ

Bees from the same hive visit about 225,000 flowers a day.

These tiny packets of clay and compost give seeds their own nutrient-rich growing medium, offering protection from the drying effects of the sun and from foraging animals such as rodents and squirrels. May your wildflowers flourish with these seeds balls!

DIRECTIONS

1. In a shallow container, thoroughly mix the compost and seeds using your hands.

2. Add the powdered clay to the compost and seed mixture. Mix it all together. Many people prefer red clay, but any clay will work. Powdered clay can be purchased from a ceramic or art supply store. (Fig. 1)

3. Now for the fun part. Pour a little water onto the clay and compost. Mix together with your hands. Continue adding water, as needed, to make a doughlike consistency. (Fig. 2)

4. Pinch off a small, marble-sized piece. Roll into a ball. If crumbly, spray with water. If sticky, add more compost. (Fig. 3)

5. Lay the seed balls on a piece of newspaper to dry for at least 24 hours.

6. Store dry seed balls in cardboard boxes or paper bags in a cool, dark, dry place. Do not use plastic.

7. Time to spread the love with the seed balls you've made. No need to bury or water them. Just toss in your yard, in ditches, or barren areas and let nature do the rest!

Fig. 1: *After combining the compost and seeds together, mix in powdered clay.*

Fig. 2: *Gradually add water to make a doughlike consistency.*

Fig. 3: *Roll into marble-sized balls.*

 TAKE IT FURTHER

- Vegetable seed balls will also work for your garden.

- Organize a seed ball–planting event in your community inspired by the work of others. In Texas, a group of one hundred bicyclists tossed 4,000 seed balls along a river trail. Dutch citizens and government entities combined efforts to sow bee-friendly plants along a 4 1/2-mile (7 km) stretch of road in the Netherlands named the "Honey Highway."

- Read the book *The One-Straw Revolution: An Introduction to Natural Farming* by Masanobu Fukuoka, who brought worldwide attention to the value of seed balls.

A BEE-FRIENDLY LANDSCAPE

YOU WILL NEED

- a sunny area
- compost
- plant starters
- seeds
- garden tools for digging (spade, trowel, shovel)
- water
- mulch
- pruning shears

BEE BUZZ

Bees love flowers that are shallow or have landing platforms.

FUN FOR KIDS

Help children learn and remember the names of the plants you have sown by making garden stakes. Write the plant names using permanent markers on rocks, plastic or wooden spoons, or vinyl slats from recycled venetian blinds.

Create a sanctuary for honeybees and other beneficial pollinators using these gardening tips.

DIRECTIONS

1. **Plan.** Begin preparations months in advance of your growing season. When deciding on a garden location, look for a spot with at least 6 hours of sunlight. If shade is your only option, take care to choose plants able to grow only in low light conditions.

2. **Prepare the soil.** Take note of the soil composition. Amend the soil with compost to improve drainage for clay soil and help maintain moisture for sandy areas. (Fig. 1)

3. **Decide what to plant.** Start with native plants and herbs. Find out what grows best in your area by consulting an extension office, gardening club, botanical garden, or local nursery specializing in native plants. The Resources section of this book offers nectar and pollen plant sources.

4. **Plant.** Use starter plants to give you a head start in establishing your garden or seeds to cover a wide area. Plant trees and shrubs in the fall to establish roots that will survive a hot, dry summer.

5. **Water.** Water plants immediately and thoroughly after planting. Provide water for wildlife and attract all kinds of insects and animals to your garden with a birdbath, a sloping pool, or even a muddy area.

6. **Mulch.** Discourage weeds and retain soil moisture by mulching with bark, leaves, or straw. (Fig. 2)

7. **Maintain.** Prune dead flowers, leaves, and stems to promote new growth. Weed, water, and watch your garden grow.

8. **Go organic.** Try to limit or eliminate pesticide use in the yard, garden, and apiary.

Fig. 1: *Compost is a great addition to any type of soil.*

Fig. 2: *Mulching decreases the need to water and weed.*

TAKE IT FURTHER

Plan for next year. Determine periods of the year when there were no blooms in your garden and identify plants to fill those gaps.

🌸 TIPS FOR BEE-FRIENDLY PLANTS

Fig. 3: *Mass plantings attract pollinators and make food collecting more convenient.*

Fig. 4: *Dandelions, often considered a weed, are a source of early spring nectar and pollen.*

Go native. Choose native plants whenever possible. Benefits include reduced water requirements and hardier plants. Avoid cultivars, which have been developed to appeal to people—not necessarily to attract pollinators. Many hybrids have exchanged showy beauty for a decrease in nectar production. When in doubt, check with your local native nursery.

Plant herbs. Bees love many herbs, including rosemary, mint, oregano, garlic, sage, and lavender. Herbs are easy to grow, convenient to use in cooking and home remedies, and smell great!

Variety is the spice of life. Different colors and flower shapes attract a wide range of pollinators.

Group same plants together. Swaths of color provide a strong visual signal for foraging bees and aid in efficient food gathering. (Fig. 3)

Vary bloom cycles. Bees need variation, so plant early bloomers like dandelion to provide pollen vital to brood buildup and fall flowers like goldenrod to help them stockpile winter stores.

Blooming weeds. Let weeds grow in wild spaces. Many wild flowering plants considered to be a nuisance are excellent pollen and nectar producers. Allow plants to reseed before cutting them down. (Fig. 4)

HONEYBEE CELEBRATION

YOU WILL NEED

- a coordinator
- a location
- volunteers
- publicity person
- tables
- supplies for specific activities

The entire town of Clarkson, Kentucky, comes out to celebrate honeybees with a parade, events, and decorated storefronts and yards as a part of the Clarkson Honeyfest.

Boost community awareness about the value of honeybees and other pollinators by organizing a free event open to the public or a private party for family and friends.

DIRECTIONS

1. Collaborate with friends and community members to organize and coordinate the event. Identify key team members to help arrange the event, identify the location, solicit a team of volunteers, publicize the event, and manage all speakers and activities.

2. Decide what kind of event it will be and then move forward with the details and logistics. Will it be a large-scale event or one single activity like a citywide scavenger hunt? A family affair or an adult audience? Will there be food? Are there space limitations and considerations? What kinds of activities would you like to have? Will you have workshops and speakers?

3. Assign volunteers to take care of food, specific activities, entertainment, and cleanup.

4. Effective publicity is absolutely essential to a successful event. Utilize the power of social media and word of mouth in addition to emails, flyers, posters, and other publicity. Contact your local newspapers and radio and television stations to arrange pre-event publicity and event-day coverage.

5. Be prepared to talk about the value and plight of pollinators.

BEE BUZZ

To make a prairie it takes
  a clover and one bee,
One clover, and a bee,
And revery.
The revery alone will do
If bees are few.

—Emily Dickinson

CELEBRATION ACTIVITIES

The possibilities for your celebration, either large or small, are endless; here are a few ideas to get you going.

Activities

- Display an observation hive.
- Supply materials such as newspaper, chenille sticks, yellow tape, etc. for people to create their own bee costumes, hats, and puppets.
- Organize a pollinator's parade. Encourage folks to decorate bikes, skateboards, and carts for a human-powered parade through a farmers' market, park, or shopping mall.
- Set up a photo booth. You supply the accessories; they snap the photos. Provide hats, antennas, wings, crowns, flowers, message posters, and other accoutrements with which people can create dramatic ensembles.
- Coordinate a silent auction. Solicit items from businesses and beekeepers that will be auctioned during the event.
- Invite local artists to create painted hives (Lab 47) to be displayed or auctioned off.

Games

- Toss the Hive Tool: Paint a target similar to a dartboard on the ground. Take turns tossing the tool and adding up the points. Award extra points if it sticks in the ground.
- Scavenger Hunt: Work with vendors at a farmers' market in advance to be part of a scavenger hunt. Give customers a card to be stamped by vendors with a list of things to find, such as comb honey, nectar plants, food made with honey, a person with a bee hat, or a specific food. Turn cards in to receive a prize.
- Citywide Scavenger Hunt: Participants are given a list of specific things to photograph, such as a honeybee, butterfly, composite flower, redbud tree, vegetable plant, herb plant, water source, and beehive.

Contests

- Have a costume contest with categories such as most creative, craziest, best use of recycled materials, or the most original group costumes.
- Display anywhere from 30 to 700 paper bees of different sizes around town or in a library or other location for people to find and count. The person with the closest number counted receives a prize.
- Organize a honey bake-off or photography contest.
- The Palisade International Honeybee Festival invites third graders to participate in a spelling bee with words that begin with the letters "be."

Children's Activities

- Provide face painting.
- Give out temporary bee tattoos.
- Cover hats, shoes, and belts with plastic flowers for roving volunteers to wear. Children make bee finger puppets using decorated paper strips (2½" × 1½", or 6.4 × 3.8 cm) rolled into tubes. Once a child finds and pollinates a "flower," the volunteer brushes glitter paint onto the child's hand to represent pollen.
- Hide little pictures of flowers and/or bees for children to find and gather.
- Beehive Pokey: Sing and do the "Hokey Pokey," changing the words to reflect bee body parts, such as antennas, wings, compound eyes, pollen baskets, and stinger.
- Possible activities in this book include honey tasting (Lab 14), comb rubbings (Lab 43), rolling beeswax candles (Lab 15), making seed balls (Lab 49), or viewing bee landing strips (Lab 26).

Food (Be aware that some event sites may have food regulations.)

- Set up a honey lemonade stand at a local event.
- Provide mead, honey beer, and wine tastings. Sell tickets to trade for a certain number of tastings.
- The Narrowsburg Honeybee Fest, in Narrowsburg, New York, offers a traditional English high tea with local honey and desserts.

HELP THE POLLINATORS

Tara Chapman maintains bees at the Sustainable Food Center in Austin, Texas.

YOU WILL NEED

- energy
- passion
- initiative

BEE BUZZ

More than 80 percent of the flowering plant species in the world rely on animal pollinators for seed production. Most of these animal pollinators are insects, although birds, and even some bats, are important pollinators, too.

As beekeepers and honeybee advocates, we are motivated to do whatever possible to benefit our often overlooked pollinators. Here are a few things you can do to help the bees.

DIRECTIONS

1. **Become a beekeeper.** Working with honeybees increases personal and community awareness about pollinators and the challenges they face.

2. **Be a mentor.** Work with a new beekeeper to help him or her gain knowledge and confidence in working with bees.

3. **Join and support beekeeping organizations.** Many local, state, national, and international organizations exist to support beekeeping and promote the benefits of honeybees.

4. **Spread the word.** Present programs at schools, libraries, senior centers, or gardening clubs. Set up an educational booth or honey lemonade stand at a farmers' market, community event, museum celebration, county fair, or yard sale.

5. **Increase knowledge.** Learn more about bees by reading books, doing research on the internet, participating in seminars and conferences, or subscribing to beekeeping publications.

6. **Be a citizen scientist.** Fill out surveys with the Bee Informed Partnership; they gather information from beekeepers to better understand how to keep healthier bees. Participate in the Great Sunflower Project by recording the number of pollinators observed in your community. Help Bumble Bee Watch collect data to track and conserve bumblebees.

7. **Support pollinators.** Help your community become a Bee City USA by following the guidelines to make the world safer for pollinators one city at a time.

8. **Provide financial support.** Financially support your favorite nonprofit organization doing work to promote research, education, and programs centered on honeybees and pollinators.

9. **Donate a hive.** Purchase a hive from a nonprofit organization such as Heifer International to help bring sustainable agriculture and commerce to people in areas with a long history of poverty.

10. **Speak out.** Write letters or sign petitions to limit pesticide use and change regulations that benefit pollinators and beekeepers.

11. **Say thank you.** When you see a person, an organization, or a government entity speaking up for pollinators, planting native habitat, or using ecologically sound practices, take a moment to thank them for their efforts.

BEEKEEPERS MAKING A DIFFERENCE

John Talbert

Toni Burnham

Bo Sterk

Sarah Red-Laird

There are multitudes of ordinary people around the world educating others about the value of honeybees through classes, presentations, events, organizations, blogs, websites, and activism. Here are a few beekeepers dedicated to supporting our tiny friends, the honeybees.

"Besides humans, honeybees are the most interesting organism in the world." —John Talbert

For the past twenty years, John Talbert has been instrumental in setting up an informational booth and coordinating about 125 volunteers for the State Fair of Texas honey display, where over a million people learn about honey and honeybees annually. Throughout the years, more than 600 people have attended John's beekeeping seminars. He has mentored teens who have gone on to be beekeeping leaders on the state and national levels in the United States. www.SabineCreekHoney.com

"There is a thing that happens when people stop being afraid of bees and start being fascinated. When you stop being afraid, you start being involved." —Toni Burnham

Toni Burnham is an extraordinary networker and tireless bee ambassador supporting bees and beekeepers in the Washington, D.C., area. She sets up beehives and mentors beekeepers at

community gardens, food banks, schoolyards, cemeteries, monasteries, backyards, community centers, and hotels. www.dcbeekeepers.org

"I work to elevate beekeepers' knowledge and production, bringing them a sustainable income, which equals better food, better health, and better education. Save the bees, feed the people." —Bo Sterk

Bo Sterk, in the red shirt, has been working with Haitian beekeeping groups through FAVACA (Florida International Volunteer Corps) and Bees Beyond Borders for ten years. His book *Simple Guide to Caribbean Beekeeping* has been published in English and Haitian Creole by Bees Beyond Borders. www.beesbeyondborders.net

"I hope by inspiring a sense of fascination, wonder, and love of honeybees in children, it will lead to understanding or even advocacy as these kids grow into our leaders of tomorrow." —Sarah Red-Laird

Sarah Red-Laird is the founder of the nonprofit Bee Girl, an organization dedicated to inspiring and empowering communities to conserve bees and their habitat. Sarah supports bees throughout the United States and the world through beekeeping classes for adults and children, public lectures, educational events, and an initiative to inform and recognize land managers who provide habitat for our bees. www.beegirl.org

GLOSSARY

Abdomen—the rear body region of a honeybee containing the honey stomach, stomach, intestines, stinger, and wax glands. It is composed of nine segments.

Antenna (plural, antennae)—segmented feelers on the head of an insect with receptors to detect smell, touch, taste, and sound.

Anther—male part of a flower that produces and contains the pollen grains.

Beebread—combination of pollen, honey, and bee glandular secretions primarily fed to brood.

Bee pollen—pollen collected by bees and stored in the hive as a protein food source.

Beeswax—produced from four pairs of glands on the underside of a worker bee's abdomen.

Brood—developing bees in the egg, larval, or pupal stages.

Brood chamber—part of the hive where the brood is reared.

Compound eye—an eye made up of thousands of tiny lenses that allow a honeybee to see ultraviolet light.

Drones—the male honeybees developed from unfertilized eggs.

Head—contains the compound eyes, ocelli, antennae, mandibles, and proboscis.

Honey—a sweet liquid made from flower nectar collected by honeybees.

Legs—a honeybee has three pairs of segmented legs used to clean the antennae, brush pollen from the hairs of the body to the pollen basket, and carry pollen and propolis.

Nectar—sweet substance secreted by the flowers of a plant that bees collect and convert into honey.

Nectar guides—patterns and UV markings on flower petals to attract pollinators.

Ocelli—three simple eyes located on top of a bee's head to sense changes in light.

Ovary—female part at the base of the flower that produces and contains the ovules.

Ovules—located in the ovary of a flower; once pollinated, ovules become seeds.

Pistil—the female part of the flower containing one or more carpels, which contain the stigma, style, and ovary.

Pollen—male reproductive cells produced by flowers and consumed by bees as a protein-rich food.

Pollen basket—flattened area on the back legs of a honeybee surrounded by long curved hairs that holds pollen and propolis.

Pollination—the transfer of pollen from the anther to the stigma of a flower.

Proboscis—strawlike tongue used to collect and consume nectar, honey, and water.

Propolis—a sticky substance produced by trees and plants.

Queen—a fully developed and mated female bee.

Stamen—male part of a flower containing the anther and filament.

Stigma—sticky surface on the uppermost part of the style where the pollen lands and germinates.

Style—narrow tube that carries germinated pollen from the stigma to the ovules (eggs) in the ovary.

Super—boxes used to store surplus honey.

Thorax—the middle body segment that supports the wings and legs.

Wax cappings—the beeswax cap used to cut off frames of honey while extracting.

Wax glands—four pairs of glands in the abdomen of a honeybee that secrete beeswax.

Wings—two sets of flat, thin wings strengthened by veins. Each set of wings attaches together during flight.

Worker bee—a female bee with undeveloped reproductive organs. Worker bees perform the daily duties of the hive.

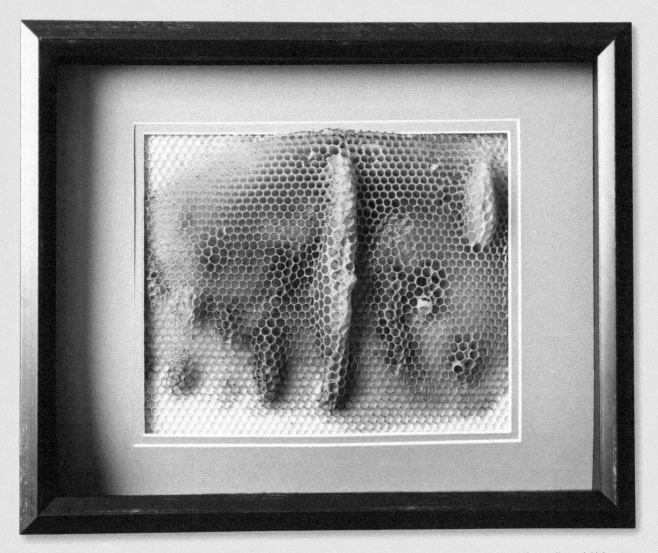

This abandoned frame of half-drawn comb was discovered while visiting a beekeeping friend. It was too beautiful to discard. Using a band saw, I cut the wooden frame and plastic foundation to fit into a standard mat board frame.

RESOURCES

BEEKEEPING AND CANDLE SUPPLIES

Australia
Pender Beekeeping Supplies
www.penders.net.au

United Kingdom
E. H. Thorne, Ltd.
www.thorne.co.uk

United States
B&B Honey Farm
www.bbhoneyfarms.com

Betterbee, Inc.
www.betterbee.com

Brushy Mountain Bee Farm
www.brushymountainbeefarm.com

Dadant & Sons
www.dadant.com

Kelley's Beekeeping
www.kelleybees.com

Mann Lake, Ltd.
www.mannlakeltd.com

CITIZEN SCIENCE PROJECTS

Bee Informed Partnership
www.beeinformed.org
*Share beekeeping information through online surveys
to help study honeybee health on a large scale.*

Bumble Bee Watch
www.bumblebeewatch.org
*Photograph bumblebees to help track and conserve
North America's bumblebees.*

The Great Sunflower Project
www.greatsunflower.org
*Record and report the number of pollinators observed
in yards and parks.*

International Bee Research Association, CSI Pollen Study
www.ibrabee.org.uk
Help measure pollen diversity in beehives throughout Europe.

National Biodiversity Data Centre
www.biodiversityireland.ie
*Participate in the Irish Pollinator Initiative by contributing
to the biodiversity database.*

ENCAUSTIC SUPPLIES

Earth Pigments
www.earthpigments.com
Powdered nontoxic pigments

TKB Trading
www.tkbtrading.com
Cosmetic dye sampler

HONEY ANALYSIS

Germany
Intertek Food Services GmbH
www.intertek.de

United Kingdom
Minerva Scientific Ltd.
www.minervascientific.co.uk

United States
Texas A&M University
vbryant@tamu.edu

HONEYBEE FESTIVALS AND CELEBRATIONS

Amherst Honey Bee Festival
www.amhersthoneybeefestival.org

Australia National Honey Month in May
www.honeybee.org.au

Clarkson Honeyfest
www.clarksonhoneyfest.com

Honeybee Fest
www.narrowsburghoneybeefest.com

Lithopolis Honeyfest
www.centralohiobeekeeper.com

NYC Honey Week
www.nychoneyweek.com

National Honey Month in September—United States
www.honey.com

National Honey Week in October—United Kingdom
www.bbka.org.uk/news_and_events/nationalhoneyweek

National Pollinator Week in June—Canada and the United States
www.pollinator.org/programs.htm

Palisade International Honeybee Festival
www.palisadehoneybeefest.org

POLLINATOR PLANTS

Lady Bird Johnson Wildflower Center
www.wildflower.org/plants
An extensive native plant database and pollinator plant list compiled for the United States and Canada.

Million Pollinator Garden Challenge
www.millionpollinatorgardens.org
Be a part of this international initiative of the National Pollinator Garden Network.

Pollinator Partnership
www.pollinator.org
Eco-regional guides for native plants in the United States and parts of Canada.

POLLINATOR SUPPORT GROUPS

Bee City USA
www.beecityusa.org
Helps communities become certified bee-friendly spaces.

Crown Bees
crownbees.com
Extensive information about Mason bees.

International Bee Research Association (IBRA)
www.ibrabee.org.uk
Promotes the value of bees by providing information on bee science and beekeeping worldwide.

Pesticide Action Network International
www.pan-international.org
This network of organizations, institutions, and individuals in more than ninety countries is working to replace the use of pesticides with ecologically sound alternatives.

Pollinator Partnership
www.pollinator.org
This organization is a pollinator support powerhouse for North America and globally.

CONTRIBUTORS

Doris Bickley (Lab 46, page 120)
Z_bick_1@yahoo.com

Wizzie Brown (Lab 40, page 106)
agrilife.org/urban-ipm

Tara Chapman (Lab 7, page 30)
www.twohiveshoney.com

Brandon Fehrenkamp (Lab 46, page 120)
www.austinbees.com

Kim Flottum (Lab 2, page 20)
www.beeculture.com

Anna Gieselman (Labs 46 and 47, pages 120 and 122)
www.beeamour.com

Ladislav Hanka (Lab 46, page 120)
www.ladislavhanka.com

Heather Isbell, Izzy's Farm & Apiary (Lab 47, page 122)
www.izzabellabeez.com

John Love (Lab 47, page 122)
lovejbl@yahoo.com

Marjorie Molnar (Lab 47, page 122)

Kristi Orcutt (Lab 5, page 26)
www.facebook.com/BrightHopeFarm

Sarah Red-Laird (Lab 26, page 74)
www.beegirl.org

Rob Rowland (Lab 42, page 110)

Mercedez Singleton (Lab 43, page 114)
www.MercedezRex.etsy.com

Bo Sterk (Lab 52, page 134)
www.beesbeyondborders.net

John Swan (Labs 7, 8, and 13, pages 30, 32, and 44)
www.WickedBeeApiary.com

Tammy West (Lab 20, page 60)
www.tammyweststudios.com

Rich Wieske and Joan Mandell
(Labs 32, 33, and 49, pages 88, 90, and 128)
www.greentoegardens.com

Tim Ziegler (Lab 42, page 110)
www.timziegler.com

PHOTOGRAPHER CREDITS

Heather Isbell uses acrylic paint protected by a polyurethane finish to create bright, vivid beehives that double as homes for bees and whimsical yard art.

ACKNOWLEDGMENTS

It takes many bees in a hive to make a sweet product. I deeply appreciate the following people and organizations who have contributed time, ideas, props, locations, kindness, and support for the making of this book: Shirley Acevedo, Linda Ayers, Robert and Lorin Bryce, Bill and Jeanine Christensen, Lisa Christensen, Kim Flottum, Allison French, Mary Guitar, Gene Helmick-Richardson, Peggy Helmick-Richardson, Rebecca Linton, Suzy McCoy, Linda Mihaly, Dr. Al Molnar, Rain Lily Farms, Margaret Bachman Reid, Helen Roberts, Cory Ryan, Mary Shaver, Jonathan Simcosky, Sunshine Community Gardens, Sustainable Food Center, Mark Wieland, Peter Williams, Lynn Wolfe, and Bill Zimmer.

ABOUT THE AUTHOR

Kim Lehman has worked for more than twenty-five years as a honeybee educator, teacher, professional storyteller, musician, and author. She has presented hundreds of programs and workshops at schools, libraries, museums, nature centers, and festivals.

As part of her children's column for *Bee Culture* magazine, Kim began the Bee Buddies club that now has more than 1,300 members in every state and two countries. She also founded the American Beekeeping Federation's Kids and Bees Program and directed this educational service about honeybees for the public at their annual conferences for seventeen years.

Besides being a hobby beekeeper with an interest in the therapeutic uses of bee products, Kim takes pleasure in gardening, travel adventures, collecting family stories, and creating with clay, metal, tile, found objects, and of course, beeswax. Visit her at www.kimlehman.com or www.beeprograms.com.

INDEX

ALSO AVAILABLE

Backyard Beekeeper

An Absolute Beginner's Guide to Keeping Bees in Your Yard and Garden

978-1-59253-919-2

Beeswax Alchemy

How to Make Your Own Soap, Candles, Balms, Creams, and Salves from the Hive

978-1-59253-979-6

The Benevolent Bee

Capture the Beauty of the Hive through Science, History, Home Remedies, and Craft

978-1-63159-286-7

CPSIA information can be obtained
at www.ICGtesting.com
Printed in the USA
LVHW071147130323
741488LV00008B/92